Global Variational Analysis
Weierstrass Integrals
on a
Riemannian Manifold

by

Marston Morse

Princeton University Press

and

University of Tokyo Press

Princeton, New Jersey

1976

Published in Japan exclusively by
University of Tokyo Press;
In other parts of the world by
Princeton University Press

Printed in the United States of America
by Princeton University Press, Princeton, New Jersey

Supported in part by Army Research Office-Durham Grant
DA-ARO-D-31-124-73-G61

Library of Congress Cataloging in Publication data will
be found on the last printed page of this book

Preface

This book is an introduction to global variational analysis of curves. It is concerned with a Weierstrass integral J on a compact differentiable manifold M_n of class C^∞ with a Reimannian structure. J is locally of the form

$$J = \int_a^b F(u(t), \dot{u}(t))dt$$

where $u = (u^1, \ldots, u^n)$ is a set of local coordinates and $\dot{u} = (\dot{u}^1, \ldots, \dot{u}^n)$, a nonvanishing set of t-derivatives. Each local preintegrand $F(u,r)$ is positive homogeneous of order 1 in its variables $r = (r^1, \ldots, r^n)$, is positive definite, and nonsingular and positive regular in senses which will be defined. The Riemannian arc length has the usual local positive definite ds^2.

The principal chapters of this book are devoted to a systematic derivation of the properties of J and its extremals, with special emphasis on a simplified definition of conjugate points. The global study is restricted to the simplest of boundary value problems, namely the search for extremals joining two prescribed points $A_1 \neq A_2$, extremals that are A_1A_2-homotopic to a prescribed curve h joining A_1 to A_2. A point pair A_1, A_2 is termed <u>nondegenerate</u> if there exists no extremal joining A_1 to A_2 on which A_2 is conjugate to A_1. Nondegenerate point pairs $A_1 \neq A_2$ are shown to be everywhere dense on $M_n \times M_n$. Degenerate point pairs can be effectively treated as limiting cases, as the last theorem of this book makes clear. Extremals reduce to geodesics on taking J as the integral of length on M_n. However, a study restricted to geodesics would be inadequate for many reasons.

Our global analysis is oriented by the general principles of critical point theory in the simplest case, namely the case in which the domain of the functional studied is in some sense nondegenerate. These principles involve the following:

I. The identification of critical elements of the functional. (Here extremals of J joining A_1 to A_2 and A_1A_2-homotopic to h.)

II. The indexing of the critical elements and the "counting" of critical elements with the same index. (Here the <u>index</u> of an extremal γ is the number of conjugate points of A_1 preceding A_2.)

III. The derivation of relations between the critical elements and relevant topological invariants.

I have recently discovered new topological invariants which are exactly what is needed in III. They are the connectivities $\underset{\sim}{R}_i$ of the pathwise components of the metric Fréchet spaces $\mathcal{F}_{A_1}^{A_2}$ associated with a point pair $A_1 \neq A_2$ on M_n. See §27. Their pathwise components are determined by the respective A_1A_2-homotopy classes but are radically different from these classes. See Morse [3] and Landis-Morse [2].

In two-thirds of the book the theorems are essentially local and classical, with a new setting and interpretation. The last third of the book comes to grips with the extremal homology relations. A final section states three global theorems for whose formulation the book has prepared. A vast field is open to the reader.

Many distinguished mathematicians have evolved or are evolving theories of global analysis or geometry. References to them will be selected on the basis of method rather than field. Relevant global studies in integral equations, minimal manifolds and differential topology will be included. Unfortunately some superlative papers will not be recognized.

I wish to acknowledge with gratitude the help which I have received from mathematicians who have been associated with me at the Institute for Advanced Study. I refer especially to Professors Dale Landis and Stewart Scott Cairns.

PART III

Minimizing arcs

Chapter 5

Necessary conditions

Chapter 6

Sufficient conditions

PART IV

Preparation for Global Theorems

Chapter 7

Elementary extremals

Chapter 8

Non-simple extremals

PART V

Global Theorems

Chapter 9

Simplifying concepts

Chapter 10

Reduction to critical point theory

INTRODUCTION

We shall give a heuristic account of some of the sections most likely to be difficult by making clear what is the main stream. §21 will be reviewed first because this is the section in which the search for extremals joining A_1 to A_2 is reduced to the search for the critical elements of a function of points. It is recommended that the reader make use of this account only after reading the first twenty sections of the book.

Summary of §21. A broken extremal γ joining A_1 to A_2 is introduced. The curve γ consists of $\nu+1$ successive elementary extremals (Def. 19.4) whose end points form a sequence

$$A_1, p_1, \ldots p_\nu, A_2 \qquad (\text{on } M_n). \tag{0.1}$$

The ν-tuple $\underset{\sim}{p} = (p_1, \ldots, p_\nu)$ is called the *vertex set* of γ. The J-length of γ is denoted by $\mathcal{J}^\nu(\underset{\sim}{p})$. A vertex set $\underset{\sim}{p}$ is an $n\nu$-tuple on the ν-fold product $(M_n)^\nu$ of M_n. Theorem 21.1 affirms the following.

(a). *The broken extremal* γ *is an extremal of* J *if and only if* $\underset{\sim}{p}$ *is a critical point of the mapping* $\underset{\sim}{p} \to \mathcal{J}^\nu(\underset{\sim}{p})$ *into* $\underset{\sim}{R}$.

In simplified form Theorem 21.2 reads as follows.

(b). *There exists an extremal* g *joining* A_1 *to* A_2 *which is* A_1A_2*-homotopic to a prescribed curve* h *joining* A_1 *to* A_2 *and which affords a minimum to* J *relative to the set of piecewise regular curves which join* A_1 *to* A_2 *and are* A_1A_2*-homotopic to* h.

Summary of §24. We pass over §22 and §23. These sections give relatively simple proofs that the definitions and theorems which have been shown to be valid in the preceding sections acquire meaning and validity when the extremal joining A_1 to A_2 is no longer simple.

Theorem 24.1 is a theorem on the measure of the set of conjugate points of a point A_1 prescribed on M_n. It implies the following.

(c). Point pairs which are conjugate to each other on no extremal joining A_1 to A_2 are everywhere dense on the product $M_n \times M_n$. (Cf. Lemma 12.1, p. 231, Morse [2].)

An extremal γ *joining a ND point pair* (A_1,A_2) *is called* singleton (Def. 24.2) if there is no other extremal joining A_1 to A_2 with the same J-length. Unless the extremals involved in Theorem 26.1 are singleton or can be made singleton by an infinitesimal alteration of the integral J, the proof of Theorem 26.1 is far less simple. That J can be replaced by an infinitesimal approximation $\hat{J}$ with the desired singleton extremals is affirmed in Replacement Lemma 24.4. The proof given in Morse [6] is restricted to the case where J is the length integral on M_n. A similar proof exists for the general J.

The existence of tractions. Appendix IV. Let f be a real-valued differentiable function with domain a topological space X. Any application of critical point theory to f requires the existence of deformations of certain subsets of X. The simplest deformations are along trajectories orthogonal to level manifolds of f. (Provided such trajectories exist.) The retracting deformations of Borsuk serve adequately in some but not in all cases. The tractions introduced in Definition 26.1 are deformations which include retracting deformations as a special case. Tractions are indispensable in the application of critical point theory to the vertex spaces $[g]_{\beta}^{\nu}$ of Definition 24.5. A function $f = f^{\nu}$ with the domain $[g]_{\beta}^{\nu}$ is defined in (26.13) with the aid of J.

The existence of tractions into the desired subspaces is established in Appendix IV. Without these existence theorems the author would not be able to prove global Theorems 26.1 and 27.1.

PART I

The Weierstrass integral J

Chapter 1

A Riemannian manifold

§1. A differentiable manifold M_n. The integrals introduced by Weierstrass in his study of the calculus of variations in parametric form (cf. Bolza [1], p. 189) had the (x,y)-plane as the underlying space. We shall replace the (x,y)-plane by an n-dimensional Riemannian manifold M_n. A Riemannian manifold is a differentiable manifold with a Riemannian structure. We begin accordingly with a brief characterization of a differentiable manifold M_n of dimension $n > 1$.

The manifold M_n is supposed of class C^∞. A priori, M_n is a connected, topological manifold, coverable by a countable union of open subspaces $\mathcal{U}, \mathcal{V}$, etc., which are the respective images of domains of C^∞-compatible presentations of these subspaces. As defined in Morse-Cairns [1], p. 29, a presentation is a homeomorphism,

$$(1.0) \qquad (\phi, U) ,$$

of an open subspace U of an n-space R^n of rectangular coordinates $u^1,\ldots,u^n$ onto an open subspace $\mathcal{U}$ of M_n.

The differential structure of M_n is determined by the set (denoted by $\mathcal{P} M_n$) of all presentations of form (1.0) which are C^∞-compatible,* in the sense of Morse-Cairns [1], p. 31, with a

*One could replace the condition of C^∞-compatibility by the condition of C^3-compatibility and deal with manifolds of class C^3. Manifolds of class C^∞ are simpler to work with and are chosen so that our major objective, an exposition of variational topology, will not be obscured by irrelevant details.

countable set* S of C^∞-presentations of open subsets of M_n, given as covering M_n. It is a basic theorem that the presentations† in $\mathcal{P}M_n$, taken arbitrarily in pairs are C^∞-compatible. A particular consequence of our hypotheses is that for each presentation $(\phi,U) \in \mathcal{P}M_n$ the restriction $\hat{\phi} = \phi|\hat{U}$ of ϕ to an open subset $\hat{U}$ of U is a presentation in $\mathcal{P}M_n$.

Local coordinates of points on M_n. Let a presentation (ϕ,U) of a subspace $\mathcal{U}$ of M_n be given. The coordinates $(u^1,\ldots,u^n)$ of a point $u \in U$ are called local coordinates of the point $\phi(u) = p \in \mathcal{U}$, or more precisely the ϕ-coordinates of p. The coordinate domain U of a presentation (ϕ,U) will be denoted by U_ϕ, when the addition of the subscript ϕ serves to remove ambiguity.

Definition 1.1. Transition diffs. Presentations (ϕ,U) and (ψ,V) such that $\phi(U) = \psi(V)$ will be termed equivalent, as well as their coordinate domains U_ϕ and V_ψ. Corresponding to such an equivalent pair of presentations, the mapping $\theta = \psi^{-1}\circ\phi$ is a C^∞-diff of U onto V called a transition diff of U onto V. The inverse of θ is a C -diff ❽ of V onto U and is termed a transition diff of V onto U.

---

*For simplicity we shall assume that the set S of presentations is locally finite in that the range $\mathcal{U}$ on M_n of each presentation in S meets at most a finite number of ranges of other presentations in S.

†The inverses of our presentations are called charts by many differential geometers.

<u>Notation</u>, <u>curves</u> $\underset{\sim}{u}$; $\phi\cdot\underset{\sim}{u}$, etc. A mapping

$$(1.1) \qquad \underset{\sim}{u} : t \to u(t) : [a,b] \to U$$

(equivalently a curve) is called <u>regular</u> if of at least class C^1 and if $\|\dot{u}(t)\|$ never vanishes. The superdot indicates differentiation as to t. If (ϕ,U) is a presentation in $\mathcal{S}^C M_n$, the mapping

$$(1.2) \qquad t \to \phi(u(t)) : [a,b] \to M_n$$

will be denoted by $\phi\cdot\underset{\sim}{u}$ and will be termed <u>regular</u> if $\underset{\sim}{u}$ is regular. If θ is a transition diff of U onto a coordinate domain V, the mapping

$$t \to \theta(u(t)) : [a,b] \to V$$

will be denoted by $\theta\cdot\underset{\sim}{u}$.

By the <u>carrier</u> $|\underset{\sim}{u}|$ of the curve $\underset{\sim}{u}$ is meant the set of points on the curve $\underset{\sim}{u}$.

<u>Contravariantly related vectors</u>. Let (ϕ,U) and (ψ,V) be equivalent presentation in $\mathcal{S}M_n$. A regular curve in U issuing from a point $u \in U$ with differentials $du^1,\ldots,du^n$ at u is mapped by the transition diff $\theta = \psi^{-1}\circ\phi$ of U onto V onto a regular curve in V issuing from the point $v = \theta(u) \in V$, with differentials $dv^1,\ldots,dv^n$ at v such that

$$(1.2)'^{*} \qquad dv^j = \frac{\partial\theta^j}{\partial u^i}(u)\, du^i \qquad (j = 1,\ldots,n).$$

One says then that, relative to the transformation from $u \in U$ to $v = \theta(u) \in V$, the vector $du^1,\ldots,du^n$ at u is contravariantly related to the vector $dv^1,\ldots,dv^n$ at $v = \theta(u)$.

More generally a vector $(r^1,\ldots,r^n) = r$ at a point $u \in U$ and a vector $(\sigma^1,\ldots,\sigma^n) = \sigma$ at a point $v = \theta(u) \in V$ are said to be contravariantly related if

$$(1.3) \qquad \sigma^j = \frac{\partial\theta^j}{\partial u^i}(u)\, r^i \qquad (j = 1,\ldots,n)$$

or equivalently, if $\Theta = \theta^{-1}$, $u = \Theta(v)$, and

$$(1.4) \qquad r^j = \frac{\partial\Theta^j}{\partial v^i}(v)\, \sigma^i \qquad (j = 1,\ldots,n).$$

It is to be noted that when vectors r and σ are contravariantly related, r is a null vector if and only if σ is a null vector.

Covariantly related vectors. We continue with a transition diff $\theta = \psi^{-1}\circ\phi$ of U onto V. A vector $(c_1,\ldots,c_n)$ at $u \in U$ and a vector $(e_1,\ldots,e_n)$ at $v = \theta(u) \in V$ such that

$$(1.5) \qquad c_j = e_i \frac{\partial\theta^i}{\partial u^j}(u) \qquad (j = 1,\ldots,n)$$

---

*Here, as elsewhere, we adopt the summation conventions of tensor algebra.

or equivalently such that

$$e_j = c_i \frac{\partial \Theta^i}{\partial v^j}(v) \qquad (\Theta = \theta^{-1}, \ (j = 1,\ldots,n)) \tag{1.6}$$

are said to be <u>covariantly</u> related. The above vectors r and σ, contravariantly related at u and $v = \theta(u)$, respectively, and the vectors c and e covariantly related at u and $v = \theta(u)$, respectively, are such that

$$r^j c_j = \sigma^j e_j \ , \tag{1.7}$$

as one readily proves with the aid of (1.4) and (1.5).

Note that the components of covariantly related vectors are indexed by subscripts while the components of contravariantly related vectors are indexed by superscripts.

<u>A differentiability convention</u>. In the following we shall often refer to differentiable mappings of <u>closed</u> domain H. This will be done only when the mapping admits a differentiable extension over an open domain extending H.

§2. M_n as Riemannian manifold. The differentiable manifold M_n of §1 can be assigned the structure of an R-manifold* in many ways, as shown in §19 of Morse-Cairns [1]. However assigned, this structure is subject to the following conditions.

With an arbitrary presentation (ϕ,U) in $\mathcal{D}M_n$ and with each n-tuple $u \in U$ there is associated a unique, positive definite, real, symmetric quadratic form

$$a_{ij}(u)\, r^i r^j \qquad (u \in U) \tag{2.1}$$

in n variables $r^1,\ldots,r^n$ subject to the condition that each mapping $u \to a_{ij}(u)$ of U into R be of class C^∞. With a second presentation $(\psi,V) \in \mathcal{D}M_n$ and with an n-tuple $v \in V$ there is similarly associated a second positive definite, symmetric, quadratic form

$$b_{ij}(v)\, \sigma^i \sigma^j \qquad (v \in V). \tag{2.2}$$

The pairs of quadratic forms so assigned to the respective pairs of presentations in $\mathcal{D}M_n$ shall satisfy two conditions of compatibility.

I. Restriction Condition. If $\hat{U}$ is a nonempty open subset $\hat{U}$ of U a restriction $(\hat{\phi},\hat{U})$ of a presentation (ϕ,U) is a presentation in $\mathcal{D}M_n$. The quadratic form associated with $(\hat{\phi},\hat{U})$

*In this context R abbreviates Riemannian.

at a point $u \in \hat{U}$ shall be the quadratic form (2.1) already associated with (ϕ, U) at the point u.

II. <u>Equivalence Condition</u>. If the presentations (ϕ, U) and (ψ, V) are <u>equivalent</u> (that is, if $\phi(U) = \psi(V)$) then for an n-tuple $u \in U$ and n-tuple $v \in V$ such that $\phi(u) = \psi(v)$, it is required that

$$(2.3) \qquad a_{ij}(u)\, r^i r^j = b_{hk}(v)\, \sigma^h \sigma^k$$

for every pair of vectors r and σ which are contravariantly related at u and v.

More generally, if $\phi(U) \cap \psi(V) \neq \emptyset$ there exist unique subsets $\hat{U}$ and $\hat{V}$, respectively, of U and V such that

$$\phi(U) \cap \psi(V) = \phi(\hat{U}) = \psi(\hat{V}) \ .$$

Condition I is applicable to $(\hat{\phi}, \hat{U})$ and $(\hat{\psi}, \hat{V})$ separately. Condition II is applicable to $(\hat{\phi}, \hat{U})$ and $(\hat{\psi}, \hat{V})$ as "equivalent" presentations.

<u>The</u>* R-<u>length</u> $L(\gamma)$ <u>of a regular curve</u> γ <u>on</u> M_n. A curve $t \to \gamma(t)$ on† M_n is termed regular if it is decomposable into successive overlapping arcs each of which is the image under some presentation (ϕ, U) of a regular arc in U. The compatibility of different presentations in $\mathcal{D}M_n$ makes it certain

---

*R here abbreviates Riemannian.

†$\gamma(t)$ is a point on M_n.

that the regularity of γ is independent of the choice of the presentations used to test this regularity.

To define the R-length of γ we begin with the case in which the carrier $|\gamma|$ of γ is included in the range $\phi(U)$ on M_n of a presentation $(\phi,U) \in \mathcal{B}M_n$. Suppose that the parameter t of γ has for domain a closed interval $[a,b]$. The image of γ under ϕ^{-1} is by hypothesis a regular curve

$$(2.4)' \qquad t \to u(t) : [a,b] \to U .$$

If (2.1) gives the Riemannian form associated with (ϕ,U), we shall assign γ the R-length

$$(2.4)'' \qquad L(\gamma) = \int_a^b [a_{ij}(u(t))\, \dot{u}^i(t)\, \dot{u}^j(t)]^{\frac{1}{2}}\, dt .$$

If the carrier $|\gamma|$ of γ is included in the range of a second presentation $(\psi,V) \in \mathcal{B}M_n$, the compatibility conditions, I and II, imply that the length of γ, defined with the aid of the second form, (2.2), equals the length $L(\gamma)$ given by (2.4)".

A regular curve γ on M_n which is compact is a finite sequence of successive regular subarcs $\gamma_1,\ldots,\gamma_r$ each of whose R-lengths is definable as in the preceding paragraph. One then sets

$$(2.5) \qquad L(\gamma) = L(\gamma_1) + \cdots + L(\gamma_r) \quad .$$

One sees that $L(\gamma)$ is independent of admissible decompositions of γ. See §4 for further details.

Piecewise regular curves on M_n. A finite sequence of regular curves on M_n joined so as to define a continuous curve on M_n is called a piecewise regular curve on M_n. The R-length of such a curve is taken as the sum of the R-lengths of the component regular curves.

A metric on M_n. A metric on M_n will now be defined. Given any two points, p and q of M_n, let $d(p,q)$ denote the G.L.B of R-lengths of piecewise regular curves on M_n which join p and q on M_n. A distance $d(p,q)$ so defined satisfies the usual three axioms on distances in a metric space. A topology on M_n, defined in the usual way in terms of this metric, will have the same open sets as does the topology given with M_n. No variational theory is required to prove these elementary facts. However the variational theory, as we shall develop it, will imply the following deeper theorem.

Theorem 2.1. *If the R-manifold M_n is compact and if ε is a sufficiently small positive constant, then any two distinct points p and q on M_n such that $d(p,q) < \varepsilon$ can be joined by one and only one geodesic which has the length $d(p,q)$. This arc* varies continuously on M_n with its end points p and q.*

---

*The term arc is reserved for simple curves; that is, for curves whose points are in one-to-one continuous correspondence with an interval of parameter values.

This theorem is a by-product of the theory of Weierstrass integrals in Chapter 7. The integral (2.4)" is admissible as a Weierstrass integral.

We shall now reformulate the compatibility condition in a way that can be extended in our conditions on Weierstrass integrals in Chapter 2. This reformulation requires the introduction of the concept of an R-preintegrand and, in Chapter 2, of the corresponding concept of a Weierstrass preintegrand.

R-preintegrands. Corresponding to the presentation (ϕ,U) and the quadratic form (2.1) associated with (ϕ,U), we shall set

$$(2.6) \qquad \mathcal{f}(u,r) = (a_{ij}(u)\, r^i r^j)^{\frac{1}{2}} \qquad ((u,r) \in U \times \dot{R}^n)$$

Here $\dot{R}^n$ denotes the space of nonnull vectors $r = (r^1,\ldots,r^n)$ at points $u \in U$, vectors which are contravariantly transformed as in (1.3). With $\mathcal{f}$ so defined, the integral (2.4)" takes the form

$$(2.7) \qquad L(\gamma) = \int_a^b \mathcal{f}(u(t),\dot{u}(t))\, dt .$$

We term $\mathcal{f}(u,r)$ the preintegrand of the R-integral (2.7). There is a unique preintegrand associated with each presentation (ϕ,U).

Corresponding to the presentation (ψ,V) we shall similarly set

(2.8) $$\mathcal{g}(v,\sigma) = (b_{ij}(v)\ \sigma^i\sigma^j)^{\frac{1}{2}}$$

for $(v,\sigma) \in V \times \dot{R}^n$ and term $\mathcal{g}$ the R-preintegrand associated with the presentation (ψ,V).

The compatibility conditions I and II can be equivalently formulated in terms of R-preintegrands as follows:

I. Restriction Condition. When $(\hat{\phi},\hat{U})$ is a presentation which is a restriction of a presentation (ϕ,U) in $\mathcal{D}M_n$, the R-preintegrands $\hat{\mathcal{f}}$ and $\mathcal{f}$ associated respectively with $(\hat{\phi},\hat{U})$ and (ϕ,U) must satisfy the condition

(2.9) $$\hat{\mathcal{f}}(u,r) = \mathcal{f}(u,r) \qquad ((u,r) \in \hat{U} \times \dot{R}^n) .$$

II. Equivalence Condition. If (ϕ,U) and (ψ,V) are equivalent presentations in $\mathcal{D}M_n$, and $\mathcal{f}$ and $\mathcal{g}$, respectively, the associated R-preintegrands then

(2.10) $$\mathcal{g}(v,\sigma) = \mathcal{f}(u,r)$$

for $u \in U$ and $v \in V$, whenever $\phi(u) = \psi(v)$ and the vector r at u is contravariantly related to the vector σ at v.

Conditions I and II are implied by the following general condition.

Compatibility Condition A. Let (ϕ,U) and (ψ,V) be arbitrary presentations in $\mathcal{D}M_n$ with $\mathcal{f}$ and $\mathcal{g}$, R-preintegrands associated respectively with (ϕ,U) and (ψ,V). If $u \in U$ and $v \in V$ are such that $\phi(u) = \psi(v)$ and if r is a nonnull n-tuple at u which is contravariantly related to an n-tuple σ at v, then $\mathcal{g}(v,\sigma) = \mathcal{f}(u,r)$. We say then that $\mathcal{g}$ and $\mathcal{f}$ are <u>compatible</u> relative to (ϕ,U) and (ψ,V).

Conversely A is implied by I and II, provided I and II are true for <u>all</u> presentations in $\mathcal{D}M_n$ for which their hypotheses are satisfied.

<u>Homogeneity of R-preintegrands</u>. R-preintegrands such as $\mathcal{f}$ and $\mathcal{g}$ satisfy a simple condition of homogeneity. This condition takes the form of identities

$$(2.11) \qquad \mathcal{f}(u,kr) \equiv k\,\mathcal{f}(u,r); \quad \mathcal{g}(v,k\sigma) \equiv k\,\mathcal{g}(v,\sigma)$$

valid for any positive number k and for (u,r) in the domain of $\mathcal{f}$ and (v,σ) in the domain of $\mathcal{g}$. In the case of $\mathcal{f}$ (and similarly in the case of $\mathcal{g}$) this homogeneity condition implies the following.

The R-length of a regular curve

$$(2.12) \qquad \underset{\sim}{u} : t \to u(t) : [a,b] \to U$$

remains unchanged if one replaces the curve $\underset{\sim}{u}$ by an admissible reparametrization

$$\underset{\sim}{w} : \tau \to w(\tau) : [c,d] \to U$$

of $\underset{\sim}{u}$ (characterized in Def. 4.1).

R-linear parametrizations. A piecewise regular curve $\underset{\sim}{u}$ of form (2.12) will be said to have an* RL-parametrization if for some value $c \in [a,b]$ and for $a \leq t \leq b$

$$(2.13) \qquad \int_c^t \mathcal{f}(u(\alpha),\dot{u}(\alpha))d\alpha \equiv pt + q ,$$

where p and q are constants of which $p > 0$. If $p = 1$, $\underset{\sim}{u}$ will be said to be R-parametrized. If $p = 1$ and $q = 0$, $\underset{\sim}{u}$ is said to be R-parametrized by R-length measured from the point $u(c)$ on $\underset{\sim}{u}$.

The following lemma is immediate. In it $\dot{u}(t)$ denotes $\frac{d}{dt} u(t)$.

Lemma 2.1. *A regular curve of form (2.12) has an RL-parametrization if and only if*

$$(2.14)' \qquad \frac{d}{dt} \mathcal{f}(u(t), \dot{u}(t)) \equiv 0 \qquad (a \leq t \leq b).$$

If $\underset{\sim}{u}$ has an RL-parametrization for which (2.13) holds, then

$$(2.14)'' \qquad p \equiv \mathcal{f}(u(t), \dot{u}(t)) \qquad (a \leq t \leq b).$$

The following lemma is useful.

Lemma 2.2. *A regular curve $\underset{\sim}{u}$ of form (2.12) admits a reparametrization,*

---

*L abbreviates linear and R Riemannian.

(2.15) $$\underset{\sim}{w} : s \to w(s) : [a',b'] \to U$$

in terms of R-length measured from a point $u(c)$ of $\underset{\sim}{u}$.

We introduce the algebraic R-length

$$\lambda(t) = \int_c^t f(u(\alpha), \dot{u}(\alpha))\, d\alpha \qquad (a \leq t \leq b)$$

measured along $\underset{\sim}{u}$ from the point $u(c)$ on $\underset{\sim}{u}$. Set $[\lambda(a),\lambda(b)] = [a',b']$. The transformation $s = \lambda(t)$ has a unique inverse $t = \mu(s)$, defined and of class C^1 for $s \in [a',b']$. Set

(2.16) $$u(\mu(s)) \equiv w(s) \qquad (s \in [a',b']).$$

The mapping

$$s \to w(s) : [a',b'] \to U$$

is a reparametrization of the curve $\underset{\sim}{u}$. Note that $\lambda(\mu(s)) \equiv s$ and

(2.17) $$\begin{aligned} f(w(s), \dot{w}(s)) &\equiv f\big(u(\mu(s)), \dot{u}(\mu(s))\big)\dot{\mu}(s) \\ &\equiv \lambda'(\mu(s))\, \dot{\mu}(s) \equiv 1 \qquad (s \in [a',b']). \end{aligned}$$

The point $u(c) = w(0)$ and

$$\int_c^s f(w(\alpha), \dot{w}(\alpha))\, d\alpha \equiv s - c \qquad (s \in [a',b']).$$

Lemma 2.2 follows.

3.1

Chapter 2

Local Weierstrass Integrals

§3. Weierstrass preintegrands. Weierstrass integrals, as we shall define them on our Reimannian manifold M_n, are generalizations of R-length on M_n. With Weierstrass, M_n was the (x,y)-plane, so that no presentation theory was required. However, general differentiable manifolds presuppose a set of compatible presentations. It is accordingly necessary to begin with a presentation (ϕ, U) in $\mathcal{D} M_n$ and replace the R-preintegrands

$$(3.0) \qquad (u,r) \to \mathcal{f}(u,r) = (a_{ij}(u)\ r^i r^j)^{\frac{1}{2}} \qquad ((u,r) \in U \times \dot{R}^n)$$

by more general preintegrands.

To this end there is associated with each presentation $(\phi, U) \in \mathcal{D} M_n$ a unique mapping

$$(3.1)^* \qquad (u,r) \to F(u,r) : U \times \dot{R}^n \to R$$

of class C^∞, subject to the homogeneity condition,

$$(3.2) \qquad F(u,kr) = kF(u,r),$$

valid for each pair (u,r) in the domain of F and for each positive constant k.

---

*The mappings F, G, etc., associated with the respective presentations in $\mathcal{D} M_n$ will be called W-preintegrands. W abbreviates Weierstrass.

There is similarly associated with a presentation (ψ,V) in $\mathcal{P}M_n$ a C^∞-mapping

$$(3.3) \qquad (v,\sigma) \to G(v,\sigma) : V \times \dot{R}^n) \to R ,$$

subject to the homogeneity condition,

$$(3.4) \qquad G(v,k\sigma) = kG(v,\sigma)$$

for each pair (v,σ) in the domain of $G(v,\sigma)$ and for each positive constant k.

Pairs of W-preintegrands F and G associated, respectively, with presentations (ϕ,U) and (ψ,V), shall satisfy conditions of <u>compatibility</u> analogous to the compatability conditions imposed on R-preintegrands. These conditions follow.

I. <u>Restriction Condition</u>. When $(\hat{\phi},\hat{U})$ and (ϕ,U) are presentations in M_n of which $(\hat{\phi},\hat{U})$ is a restriction of (ϕ,U), the W-preintegrands $\hat{F}$ and F, associated, respectively, with $(\hat{\phi},\hat{U})$ and (ϕ,U), shall satisfy the condition

$$(3.5)' \qquad \hat{F}(u,r) = F(u,r) \qquad ((u,r) \in \hat{U} \times \dot{R}^n) .$$

II. <u>Equivalence Condition</u>. If (ϕ,U) and (ψ,V) are equivalent presentations in $\mathcal{P}M_n$ and F and G, respectively, the associated W-preintegrands, then

$$(3.5)'' \qquad G(v,\sigma) = F(u,r)$$

for $u \in U$ and $v \in V$, whenever $\phi(u) = \psi(v)$ and the vector r at u is contravariantly related to the vector σ at v.

Weierstrass integrals associated with the respective presentations (ϕ,U), (ψ,V), etc. in $\mathcal{E}M_n$ will now be defined.

The W-integrals J_F, J_G, etc. By hypotheses F and G are preintegrands associated with the presentations (ϕ,U) and (ψ,V) respectively. Corresponding to a piecewise regular curve,

(3.6)' $$\underset{\sim}{u} : t \to u(t) : [a,b] \to U,$$

in U, we set

(3.6)" $$J_F(\underset{\sim}{u}) = \int_a^b F(u(t),\ \dot{u}(t))\ dt\ .$$

Corresponding to a piecewise regular curve

(3.7)' $$\underset{\sim}{v} : t \to v(t) : [a,b] \to V$$

in V we similarly set

(3.7)" $$J_G(\underset{\sim}{v}) = \int_a^b G(v(t),\ \dot{v}(t))\ dt.$$

The following lemma is a consequence of the compatability condition (3.5)".

Lemma 3.1. Let (ϕ,U) and (ψ,V) be equivalent presentations with which preintegrands F and G are respectively associated.

<u>Set</u> $\theta = \psi^{-1}\circ\phi$. <u>If</u> $\underset{\sim}{u}$ <u>is the regular curve</u> (3.6)', <u>and</u>*
$\underset{\sim}{v} = \theta \cdot u$ <u>then</u>

$$(3.8) \qquad \int_a^b F(u(t), \dot{u}(t))\, dt = \int_a^b G(v(t), \dot{v}(t))\, dt .$$

<u>The</u> W-<u>integral</u> J <u>on</u> M_n. $J(\gamma)$ will be defined for each piecewise regular curve γ on $\overset{\circ}{M}_n$, beginning with the case in which γ is regular.

$J(\gamma)$, <u>when</u> γ <u>is regular</u>. Suppose first that the carrier $|\gamma|$ of γ is included in some coordinate range. In this case $\phi^{-1}\cdot\gamma$ is a regular curve $\underset{\sim}{u}$ of the form (3.6)'. We then set $J(\gamma) = J_F(\underset{\sim}{u})$. More generally, a regular curve γ on M_n can be decomposed into a finite sequence of regular curves γ_i, $i = 1,\ldots,r$ such that for each i, some presentation (ϕ_i, U_i) exists in $\mathcal{D} M_n$ such that $|\gamma_i| \subset \phi_i(U_i)$. $J(\gamma_i)$ is then well-defined and we set

$$(3.9) \qquad J(\gamma) = J(\gamma_i) + \cdots + J(\gamma_r) .$$

One can justify the definition (3.9) of $J(\gamma)$ by showing that the value of $J(\gamma)$ is independent of the way γ is decomposed into a sequence of regular curves γ_i, and of the choice of presentations (ϕ_i, U_i) in whose ranges $\phi_i(U_i)$ the respective carriers $|\gamma_i|$ are included.

When γ is a sequence $\eta_1,\ldots\eta_\mu$ of regular curves joined to make a piecewise regular curve γ one set

---

*We write $\underset{\sim}{v} = \theta\cdot\underset{\sim}{u}$ if $v(t) = \theta(u(t))$.

$$J(\gamma) = J(\eta_1) + \cdots + J(\eta_\mu) . \tag{3.10}$$

The Compatibility Conditions I and II on W-preintegrands are implied by the following analogue of Compatibility Condition A of §2.

Definition 3.1. *Compatibility Condition* B. Let (ϕ,U) and (ψ,V) be two presentations in ϑM_n and let F and G be two W-preintegrands associated respectively with (ϕ,U) and (ψ,V). Then F and G will be understood to be mutually *compatible relative to* (ϕ,U) *and* (ψ,V) if Compatibility Condition A of §2 is satisfied with $f(u,r)$ and $g(v,\sigma)$ replaced by $F(u,r)$ and $G(v,\sigma)$, respectively.

With relative compatibility of W-presentations understood in this sense the following theorem can be stated.

Theorem 3.1. *Let* K *be a subset of the presentations of* ϑM_n *whose ranges cover* M_n. *With the presentations in* K *let there be associated W-preintegrands which are pairwise mutually compatible.*

There then exists a unique set of W-preintegrands associated with the respective presentations in ϑM_n *which are pairwise mutually compatible and which include the W-preintegrands already associated with the presentations in* K.

The proof of Theorem 3.1 is formally similar to the proof of Theorem 19.7 of Morse-Cairns [1], with obvious changes of notation.

The following definition is useful.

Definition 3.2. *A W-structure on our Riemannian manifold* M_n. The manifold M_n will be said to have a *W-structure* if mutually compatible homogeneous W-preintegrands F have been associated with each presentation in $\mathcal{S}M_n$, together with the corresponding W-integrals J_F (termed local) and in terms of these local integrals a W-integral J (termed global) defined as above on each admissible* curve γ on M_n.

R-preintegrands are special cases of W-preintegrands. However, a W-preintegrand can be radically different from an R-preintegrand as the following example shows.

Example 3.1. Suppose that M_n is an open subspace of a 2-plane with coordinates u^1,u^2. A vector r is a nonnull 2-tuple (r^1,r^2). Let $\|r\|$ be the norm of r. We introduce a preintegrand

$$F(u^1,u^2,r^1,r^2) = u^1r^2 - u^2r^1 + A(u^1,u^2)\ \|r\|$$

where the mapping $(u^1,u^2) \to A(u^1,u^2)$ is real analytic with $A(u^1,u^2) > 0$ in an open subset of the (u^1,u^2)-plane. Integrals of the form $\int_a^b F(u,u')\,dt$ occur in the restricted problem of three bodies. Cf. G. D. Birkhoff [1], p. 11.

This example shows that $F(u,r)$ may not equal $F(u,-r)$. However, if $\mathcal{F}$ is defined as in (2.6), $\mathcal{F}(u,-r) = \mathcal{F}(u,r)$.

---

*For the present admissible curves on M_n are the piecewise regular curves on M_n.

§4. The homogeneity of W-preintegrands. The homogeneity condition

(4.1) $$F(u,kr) \equiv kF(u,r) \qquad \text{(cf. (3.2))}$$

on the W-preintegrand F is satisfied for $(u,r) \in U \times \dot{R}^n$ and for each positive k. For these values of (u,r) and k, the following relations are implied:

(4.2) $$r^i F_{r^i}(u,kr) = F(u,r)$$

(4.3) $$F_{r^i}(u,kr) = F_{r^i}(u,r) \qquad (i = 1,\ldots,n)$$

(4.4) $$r^j F_{r^i r^j}(u,kr) = 0 \qquad (i = 1,\ldots,n)$$

(4.5) $$r^i F_{r^i u^j}(u,kr) = F_{u^j}(u,r) \qquad (i = 1,\ldots,n)$$

(4.6) $$F_{u^i}(u,kr) = kF_{u^i}(u,r) \qquad (i = 1,\ldots,n)$$

(4.7) $$kF_{r^i r^j}(u,kr) = F_{r^i r^j}(u,r) \qquad (i,j = 1,\ldots,n)$$

Proof of the homogeneity relations. Differentiating the members of (4.1) as to k, yields (4.2). Differentiating the members of (4.1) as to r^i, yields (4.3). Differentiating the members of (4.3) as to k, yields (4.4). On differentiating the members of (4.2) as to u^j, (4.5) follows. Differentiating the members of (4.1) as to u^i, gives (4.6) and differentiating the members of (4.3) as to r^j, gives (4.7). From (4.4) we infer that the n-square determinant

$$|F_{r^i r^j}(u,r)| \equiv 0 \quad ((u,r) \in U \times \dot{R}^n) . \tag{4.8}$$

Relations similar to the above hold when the W-preintegrand G replaces F. One starts with the relation (3.4). When G replaces F, (v,σ) replaces (u,r).

The following definition sharpens the concept of change of parameter on a curve.

Definition 4.1. An admissible change of parameter. Let a presentation (ϕ,U) be given, together with a regular mapping (equivalently a regular curve)

$$\underset{\sim}{u} : t \to u(t) : [a,b] \to U . \tag{4.9}$$

A surjective homeomorphism

$$\tau \to \eta(\tau) : [c,d] \to [a,b] \tag{4.10'}$$

of class C^1 such that $\dot{\eta}(\tau) > 0$ yields a regular reparametrization

$$\underset{\sim}{w} : \tau \to w(\tau) : [c,d] \to U \tag{4.10''}$$

of the curve (4.9) in which $w(\tau) \equiv u(\eta(\tau))$.

The preceding definition enters explicitly in the following lemma. The superscript τ or t on F or its partial derivatives means evaluation for $(u,r) = (w(\tau),\ \dot{w}(\tau))$ or $(u,r) = (u(t),\ \dot{u}(t))$, respectively.

Lemma 4.1. *If under the conditions of the preceding definition the derivative of* $F^t_{r^i}$ *as to* t *exists for* $a \leq t \leq b$ *and* $i = 1,\dots,n$, *the derivative of* $F^\tau_{r^i}$ *as to* τ *then exists for* $c \leq \tau \leq d$ *and the* n *relations*

$$\frac{d}{d\tau} F^\tau_{r^i} - F^\tau_{u^i} \equiv \left[\frac{d}{dt} F^t_{r^i} - F^t_{u^i}\right] \dot{\eta}(\tau) \qquad (i - 1,\dots,n) \tag{4.11}$$

hold for $c \leq \tau \leq d$, *subject to the condition* $t = \eta(\tau)$.

Subject to the condition $t = \eta(\tau)$, $w(\tau) = u(t)$ and $\dot{w}(\tau) = \dot{u}(t)\ \dot{\eta}(\tau)$. It follows from the homogeneity conditions (4.3) and (4.6) that, subject to the condition $t = \eta(\tau)$,

$$F_{r^i}(u(t),\ \dot{u}(t)) = F_{r^i}(w(\tau),\ \dot{w}(\tau)) \tag{4.12}$$

and

$$\dot{\eta}(\tau)\ F_{u^i}(u(t),\ \dot{u}(t)) = F_{u^i}(w(\tau),\ \dot{w}(\tau)). \tag{4.13}$$

Lemma 4.1 follows.

The following lemma is used in reducing the Euler equations in parametric form to nonparametric form. The lemma is concerned with the W-preintegrand F associated with the presentation (ϕ, U). A regular mapping

$$t \to u(t) : [a,b] \to U \tag{4.14}$$

of class C^2 is given.

Lemma 4.2. *Given the C^2-mapping (4.14), then for $a \leq t \leq b$ and for $i = 1,\ldots,n$,*

$$\dot{u}^i(t) \left[\frac{d}{dt} F^t_{r^i} - F^t_{u^i}\right] \equiv 0 , \tag{4.15}$$

where the superscript t indicates evaluation for $(u,r) = (u(t), \dot{u}(t))$.

The numerical value of the bracket in (4.15) is

$$F^t_{r^i r^j} \ddot{u}^j(t) + F^t_{r^i u^j} \dot{u}^j(t) - F^t_{u^i} \qquad (i = 1,\ldots,n) \tag{4.16}$$

On making use of the homogeneity relations (4.4) and (4.5), one confirms the truth of (4.15).

For the mapping (4.14) and for the R-preintegrand $\mathcal{f}$ the identity

$$\dot{u}^i(t) \left[\frac{d}{dt} \mathcal{f}^t_{r^i} - \mathcal{f}^t_{u^i}\right] \equiv 0 \qquad [a \leq t \leq b] \tag{4.17}$$

holds similarly.

§5. <u>The compatibility of W-preintegrands</u>. Let (ϕ,U) and (ψ,V) be "equivalent"* presentations in $\mathcal{L}M_n$. Let γ be a regular curve on M_n with carrier on $\phi(U) = \psi(V)$. Then $\phi^{-1}\cdot\gamma$ is a regular mapping

(5.1) $$\underset{\sim}{u} : t \to u(t) : [a,b] \to U$$

into U, and $\psi^{-1}\cdot\gamma$ a regular mapping

(5.2) $$\underset{\sim}{v} : t \to v(t) : [a,b] \to V .$$

<u>Theorem</u> 5.1. <u>Let F and G be preintegrands, associated, respectively, with equivalent* presentations (ϕ,U) and (ψ,V). Set $\theta = \psi^{-1}\circ\phi$. If the curves $\underset{\sim}{u}$: (5.1) and $\underset{\sim}{v}$: (5.2) are images in U and V under ϕ^{-1} and ψ^{-1} of a regular curve γ on M_n of class† C^2, then for $a \le t \le b$ and $i = 1,\ldots,n$</u>

(5.3) $$\left[\frac{d}{dt} F_{r^i}(u(t), \dot{u}(t)) - F_{u^i}(u(t), \dot{u}(t))\right] \equiv \left[\frac{d}{dt} G_{\sigma^j}(v(t), \dot{v}(t)) - G_{v^j}(v(t), \dot{v}(t))\right] \frac{\delta\theta^j}{\delta u^i}(u(t)) .$$

---

*That is with $\phi(U) = \psi(V)$.

†Theorem 5.1 holds under the assumption that γ is of class C^1 provided $\frac{d}{dt} F_{r^i}(u(t), \dot{u}(t))$ exists for $t \in [a,b]$ or, equivalently, if $\frac{d}{dt} G_{\sigma^i}(v(t), \dot{v}(t))$ exists for $t \in [a,b]$.

Under the hypotheses of the theorem the compatibility condition (3.5)" reduces to an identity

(5.4) $$F(u,r) = G(v,\sigma) \qquad ((u,r) \in U \times \dot{R}^n)$$

subject to the conditions,

(5.5)' $$v^i = \theta^i(u), \quad \sigma^h = \frac{\delta\theta^h(u)}{\delta u^k} r^k$$

or equivalently

(5.5)" $$u^i = \Theta^i(v), \quad r^h = \frac{\delta\Theta^h}{\delta v^k}(v)\, \sigma^k$$

where i, h, k have the range $1,\ldots,n$.

From (5.4), subject to the conditions (5.5), it follows that, for $i = 1,\ldots,n$,

(5.6) $$F_{r^i}(u,r) = G_{\sigma^h}(v,\sigma) \frac{\delta\theta^h}{\delta u^i}(u)$$

(5.7) $$F_{u^i}(u,r) = G_{v^h}(v,\sigma) \frac{\delta\theta^h}{\delta u^i}(u) + G_{\sigma^h}(v,\sigma) \frac{\delta^2\theta^h(u)}{\delta u^k \delta u^i} r^k$$

Now $v(t) \equiv r(u(t))$, while the vector $\dot{v}(t)$ at $v(t)$ is contravariantly related to $\dot{u}(t)$ at $u(t)$. Relations (5.6) and (5.7) accordingly hold with (u,r,v,σ) replaced by $(u(t), \dot{u}(t), v(t), \dot{v}(t))$ for $a \leq t \leq b$.

From (5.6) and (5.7) so evaluated, it follows that the left member of (5.3) equals

$$(5.8)\qquad \frac{d}{dt}\left(G_{\sigma^h}\frac{\delta\theta^h}{\delta u^i}\right) - \left(G_{v^h}\frac{\delta\theta^h}{\delta u^i} + G_{\sigma^h}\frac{\delta^2\theta^h}{\delta u^k\delta u^i}\dot{u}^k(t)\right) ,$$

where the partial derivatives of G are evaluated for $(v,\sigma) = (v(t), \dot{v}(t))$ and the partial derivatives of θ^h are evaluated for $u = u(t)$. The value (5.8) reduces to

$$(5.9)\qquad \frac{\delta\theta^h}{\delta u^i}\left(\frac{d}{dt}G_{\sigma^h} - G_{v^h}\right) + G_{\sigma^h}\left[\frac{d}{dt}\frac{\delta\theta^h}{\delta u^i} - \frac{\delta^2\theta^h}{\delta u^k\delta u^i}\dot{u}^k(t)\right]$$

where the evaluation in (5.9) is as in (5.8). So evaluated, the bracket in (5.9) vanishes.

The left member of (5.3) thus equals the first term in (5.9), thereby establishing (5.3).

<u>Corollary</u> 5.1. <u>Under the conditions of Theorem</u> 5.1 <u>the mapping</u> $\underset{\sim}{u}$: (5.1) <u>satisfies the Euler equations of</u> J_F <u>if and only if the mapping</u> $\underset{\sim}{v}$: (5.2) <u>satisfies the Euler equations of</u> J_G.

<u>Three invariants</u>. We continue with equivalent presentations (ϕ,U) and (ψ,V) and the transition diff $\theta = \psi^{-1}\circ\phi$ of U onto V. Each of the three invariants which are here introduced, is defined for each presentation in $\mathcal{D}M_n$ and has values in R. When two presentations in $\mathcal{D}M_n$ are equivalent the associated invariants have equal values under conditions which we shall presently define.

The first of these invariants is presented in Lemma 5.1 and is an entity whose vanishing defines the classical transversality condition. The second of these invariants is presented in Lemma 5.2. It is a quadratic form used in §14 to define the condition of positive regularity of a W-preintegrand F at each 2n-tuple (u,r) in the domain of F. The third of these invariants is a value of the Weierstrass $\mathcal{E}_F$-function here associated with each preintegrand F.

Introduction to Lemma 5.1. A vector λ at a point $u \in U$ and a vector μ at a point $v = \theta(u) \in V$ are contravariantly related if

$$\mu^i = \frac{\delta\theta^i(u)}{\delta u^k}\lambda^k \qquad (i = 1,\ldots,n) . \tag{5.10}$$

Let F and G be W-preintegrands associated, respectively, with the presentations (ϕ,U) and (ψ,V). The first of our lemmas on invariants follows.

Lemma 5.1. *If a 2n-tuple (u,r) in the domain of F and a 2n-tuple (v,σ) in the domain of G are related as in (5.5) and if the vector λ at u is contravariantly related to a vector μ at $v = \theta(u)$, then*

$$F_{r^i}(u,r)\lambda^i = G_{\sigma^i}(v,\sigma)\mu^i . \tag{5.11}$$

Let $v \to \Theta(v)$ be the inverse of the transition diff θ. The relation (5.6) of covariance is equivalent to the relation

(5.12) $$G_{\sigma^i}(v,\sigma) = F_{r^h}(u,r) \frac{\partial\Theta^h(v)}{\partial v^i}$$

provided (u,r) and (v,σ) are related as in (5.5). One can reduce the right member of (5.11) to the left by making use of (5.12), (5.10) and the identity

(5.13) $$\frac{\partial\Theta^h(v)}{\partial v^j} \frac{\partial\theta^j(u)}{\partial u^k} \equiv \delta^k_h , \qquad (u \in U)$$

valid when $v = \theta(u)$.

<u>Introduction to Lemma</u> 5.2. Subject to the conditions (5.5)" one infers from (5.12) that

(5.14) $$G_{\sigma^i\sigma^j}(v,\sigma) = F_{r^h r^k}(u,r) \frac{\partial\Theta^h(v)}{\partial v^i} \frac{\partial\Theta^k(v)}{\partial v^j} \quad (h,k = 1,\ldots,n).$$

The relations (5.14) imply that the matrices

$$\|F_{r^i r^j}(u,r)\|, \quad \|G_{\sigma^i\sigma^j}(v,\sigma)\|$$

are related as <u>covariant</u> tensors of the second order. See Eisenhart [1], pages 6-9. The relations (5.14), (5.10) and (5.13), together imply the truth of the following lemma.

<u>Lemma</u> 5.2. <u>Under the conditions of Lemma</u> 5.1

(5.15) $$F_{r^i r^j}(u,r)\lambda^i\lambda^j = G_{\sigma^i\sigma^j}(v,\sigma)\mu^i\mu^j .$$

<u>Definition</u> 5.1. <u>The Weierstrass functions</u> $\mathcal{E}_F$, $\mathcal{E}_G$, etc. Corresponding to a W-preintegrand F associated with a presentation (ϕ,U), $\mathcal{E}_F$ will be defined by setting

$$\mathcal{E}_F(u,\rho,r) = F(u,r) - r^i F_{r^i}(u,\rho) \tag{5.16}$$

for each triple $(u,\rho,r) \in U \times \dot{R}^n \times \dot{R}^n$.

Corresponding similarly to a W-preintegrand G, associated with the presentation (ψ,V), $\mathcal{E}_G$ will be defined by setting

$$\mathcal{E}_G(v,\tau,\sigma) = G(v,\sigma) - \sigma^i G_{\sigma^i}(v,\tau) \tag{5.17}$$

for each triple $(v,\tau,\sigma) \in V \times \dot{R}^n \times \dot{R}^n$.

It is to be noted that $\mathcal{E}_F(u,\rho,\rho) = 0$ and $\mathcal{E}_G(v,\tau,\tau) = 0$ by virtue of the homogeneity condition (4.2).

The values of these Weierstrass $\mathcal{E}$-functions are invariant in the sense of the following lemma.

<u>Lemma</u> 5.3. <u>If</u> $\mathcal{E}_F$ <u>and</u> $\mathcal{E}_G$ <u>are associated with equivalent presentations</u> (ϕ,U), (ψ,V) <u>then if</u> $\theta = \psi^{-1}\circ\phi$,

$$\mathcal{E}_F(u,\rho,r) = \mathcal{E}_G(v,\tau,\rho) \tag{5.18}$$

<u>provided</u> $v = \theta(u)$ <u>and the</u> n-<u>tuples</u> τ <u>and</u> σ <u>at</u> v <u>are contravariant images at</u> v <u>of</u> ρ <u>and</u> r <u>respectively</u>.

Since the 2n-tuples (u,r) and (v,σ) satisfy the conditions (5.5), $F(u,r) = G(v,\sigma)$ by virtue of the compatibility condition

(3.5)". Since τ is the contravariant image at v of ρ at u

$$r^i F_{r^i}(u,\rho) = \sigma^i G_{\sigma^i}(v,\tau) \tag{5.19}$$

by virtue of Lemma 5.1. The relation (5.18) follows.

The homogeneity of $\mathcal{E}_F$ and $\mathcal{E}_G$. For (u,ρ,r) in the domain of $\mathcal{E}_F$ and for nonnull positive scalars κ and k

$$\mathcal{E}_F(u,\kappa\rho,kr) = k\mathcal{E}_F(u,\rho,r) \tag{5.20}$$

as a consequence of the homogeneity conditions (4.1) and (4.3). $\mathcal{E}_G$ satisifes similar homogeneity relations.

Preparation for Lemma 5.4. Lemma 5.4 is of basic importance in finding conditions on F and triples (u,ρ,r) in the domain of $\mathcal{E}_F$ for which $\mathcal{E}_F(u,\rho,r) > 0$. Final results will be given after the condition of positive regularity of F at 2n-tuples (u,r) in its domain has been defined and analyzed. See §14. We begin with the formula

$$\mathcal{E}_F(u,\rho,r) = F(u,r) - F(u,\rho) - (r^i - \rho^i)F_{r^i}(u,\rho) , \tag{5.21}$$

valid for each triple (u,ρ,r) in the domain of $\mathcal{E}_F$. The formula (5.21) follows from the definition (5.16) and the relation (4.2).

Lemma 5.4. *For triples* (u,ρ,r) *in the domain of* $\mathcal{E}_F$ *such that* ρ *and* r *are linearly independent,*

$$(5.22) \qquad \mathcal{E}_F(u,\rho,r) = \tfrac{1}{2}(r^i-\rho^i)(r^j-\rho^j)F_{r^ir^j}(u,X)$$

where $X = \rho + \theta(r-\rho)$ *for some value* θ *in the interval* $(0,1)$.

The right member of (5.22) is the remainder of second order in Taylor's development of $F(u,r)$ about $r = \rho$ for fixed $u \in U$. The formula (5.22) is valid because no n-tuple X of the form $\rho + \theta(r-\rho)$ vanishes for $0 < \theta < 1$ when r and ρ are linearly independent.

Part II

The Euler Equations

Chapter 3

The Euler-Riemann Equations

§6. The Weierstrass nonsingularity condition. Let F be the W-preintegrand associated with a presentation (ϕ, U) in $\mathcal{DM}_n$. As given in (3.1), F has values $F(u, r)$ and domain $U \times \dot{R}^n$. The associated Euler equations

(6.1) $$\frac{d}{dt} F_{r^i}(u, \dot{u}) - F_{u^i}(u, \dot{u}) = 0 \qquad (i = 1, \ldots, n)$$

can be understood in two senses. Formally the conditions can be regarded as conditions on a regular C^2-mapping $t \to u(t) : [a, b] \to U$ of the form

(6.2) $$F_{r^i r^j}(u, \dot{u})\ddot{u}^j - F_{r^i u^j}(u, \dot{u})\dot{u}^j - F_{u^i}(u, \dot{u}) = 0 \qquad (i = 1, \ldots, n).$$

According to (4.8) the determinant

(6.3) $$|F_{r^i r^j}(u, r)| \equiv 0 \qquad ((u, r) \in U \times \dot{R}^n)$$

so that Cramer's rule cannot be used to solve the equations (6.2) for the n-tuples $(\ddot{u}^1, \ldots, \ddot{u}^n)$.

For this reason we shall regard the Euler equations a priori, as conditions on a regular C^1-mapping $t \to u(t) : [a, b] \to U$ adjoined to the condition that the derivative as to t of $F_{r^i}(u(t), \dot{u}(t))$ exists for $i = 1, \ldots, n$ and for $a \leq t \leq b$. The existence of solutions of the Euler equation in this sense is based, as we shall see, on the assumption that the rank of the n-square matrix

(6.4) $$\| F_{r^i r^j}(u, r) \|$$

is $n-1$ for each pair $(u, r) \in U \times \dot{R}^n$. We codify this assumption as follows.

Definition 6.1. We say that F is nonsingular if the matrix (6.4) has the rank n-1 at each pair $(u, r) \in U \times \dot{R}^n$.

There is a classical condition which is equivalent to the above condition of non-singularity of F. To formulate this condition one borders the matrix (6.4) on the right by a column $x_1, \ldots, x_n, 0$ and below by a row $y_1, \ldots, y_n, 0$, where x_i and y_i are in R, obtaining thereby an (n+1)-square determinant

$$(6.5) \qquad \begin{vmatrix} k_{11} & \cdots & k_{1n}, & x_1 \\ \cdot & \cdot & \cdot & \cdot \\ \cdot & \cdot & \cdot & \cdot \\ k_{n1} & \cdots & k_{nn}, & x_n \\ y_1 & \cdots & y_n, & 0 \end{vmatrix} = D(u, r : x, y)$$

where k_{ij} denotes the ij-th element of the matrix (6.4). A basic lemma follows.

Lemma 6.1. *The rank of the matrix* (6.4) *at a pair* $(u, r) \in U \times \dot{R}^n$ *is* n-1, *if and only if* $D(u, r : r, r) \neq 0$.

The proof of this lemma is relatively simple once the relation (6.7) below has been established.

Proof of (6.7). If $A^{ij}(u, r)$ is the cofactor of the ij-th element in the determinant (6.3) then (see Bôcher [1], page 161)

$$(6.6) \qquad D(u, r : x, y) = -A^{ij}(u, r)x_i y_j \qquad ((u, r) \in U \times \dot{R}^n)$$

With (u, r) fixed, let k be an integer in the range $1, \ldots, n$ such that $r^k \neq 0$. If the first n rows of the determinant (6.5) are multiplied, respectively, by $r^1, \ldots, r^n$, added and substituted for the k-th row, the elements in the new k-th row will all be zero by virtue of (4.4), including the element in the last column, provided $r^i x_i = 0$. For fixed (u, r) the determinant (6.5) thus vanishes whenever $r^i x_i$ vanishes. For fixed (u, r) the determinant (6.5) is a bilinear form in the n-tuples x and y. Hence for fixed (u, r), $r^i x_i$ is a factor of $D(u, r : x, y)$.

By operating similarly on the columns of $D(u, r : x, y)$ we infer that for fixed (u, r), $r^i y_i$ is a factor of $D(u, r : x, y)$. Hence for (u, r) fixed in $U \times \dot{R}^n$

$$D(u, r : x, y) = -F_1(u, r) r^i x_i r^j y_j \tag{6.7}$$

introducing the Weierstrass coefficient $F_1(u, r)$ for each pair $(u, r) \in U \times \dot{R}^n$.

Completion of proof of Lemma 6.1. If the rank of the matrix (6.4) is n-1 the bilinear form (6.6) does not vanish identically. In (6.7) then, $F_1(u, r) \neq 0$ for each pair $(u, r) \in U \times \dot{R}^n$. It follows from (6.7) that $D(u, r : r, r) \neq 0$ in this case.

Conversely if $D(u, r : r, r) \neq 0$ the right and hence the left members of (6.7) do not vanish identically in the variables x_i and y_i. Hence in (6.6) at least one of the cofactors $A^{ij}(u, r)$ fails to vanish, so that the rank of the matrix (6.4) must be n-1.

Thus Lemma 6.1 is true.

The Weierstrass coefficient $F_1(u, r)$. This coefficient was introduced by Weierstrass when $n = 2$. For $n > 2$, cf. Hadamard [1], p. 95, Morse [2], p. 112 and Bliss [1], Theorem 43.1, p. 107.

The left member of (6.6) equals the left member of (6.7), so that for each $(u, r) \in U \times \dot{R}^n$ and arbitrary n-tuples x and y

$$A^{ij}(u, r)x_i y_j \equiv F_1(u, r)(r^i x_i)(r^j y_j) . \tag{6.8}$$

Regarded as an identity in the n-tuples x and y (6.8) implies that for each 2n-tuple (u, r) in the domain of F

$$A^{ij}(u, r) = F_1(u, r) r^i r^j \qquad (i, j = 1, \ldots, n). \tag{6.9}$$

From (6.9) we infer the following.

Theorem 6.1. *The rank of the matrix* $\| F_{r^i r^j}(u, r) \|$ *is* $n-1$ *if and only if* $F_1(u, r) \neq 0$.

Corresponding to the W-preintegrand G associated with the presentation (ψ, V), a W-coefficient $G_1(v, \sigma)$ is defined as was $F_1(u, r)$, on replacing (u, r) by (v, σ), and has similar properties.

From (6.9) one infers that

$$F_1(u, r) = \frac{A^{11}(u, r) + \cdots + A^{nn}(u, r)}{r^1 r^1 + \cdots + r^n r^n} \qquad ((u, r) \in U \times \dot{R}^n). \tag{6.10}$$

This formula implies that F_1 is of class C^∞ on its domain.

Invariance of the singularity condition. The basic lemma follows.

Lemma 6.2. *If* F *and* G *are W-preintegrands associated with equivalent presentations* (ϕ, U) *and* (ψ, V), *and if the* 2n-*tuples* (u, r) *and* (v, σ) *are related as in* (5.5), *then the matrices*

$$\| F_{r^i r^j}(u, r) \|, \quad \| G_{\sigma^i \sigma^j}(v, \sigma) \| \tag{6.11}$$

have the same rank, a rank $n-1$ *or less.*

This lemma is an immediate consequence of the relations (5.14) and the nonsingularity of the Jacobian matrix of the transition diff $\theta = \psi^{-1} \circ \phi$.

A corollary of this lemma and of Theorem 6.1 is that when (u, r) and (v, σ) are related, as in (5.5), $F_1(u, r) = 0$ if and only if $G_1(v, \sigma) = 0$.

The question arises, with what generality is the nonsingularity condition (Def. 6.1) on a W-preintegrand F satisfied at pairs (u, r) in the domain of F? The following theorem answers this question for R-preintegrands.

Theorem 6.2. *Each R-preintegrand $\mathcal{f}$ is nonsingular in the sense of Def. 6.1 at each pair in its domain.**

Corresponding to a presentation $(\phi, U) \in \mathcal{D}M_n$ a typical R-preintegrand $\mathcal{f}$ is defined in (2.6). Let a pair $(u_0, r_0) \in U \times \dot{R}^n$ be prescribed and fixed. To establish the theorem we must show that whatever the choice of (u_0, r_0) may be

$$\text{rank}\| \mathcal{f}_{r^i r^j}(u_0, r_0) \| = n-1 . \tag{6.12}$$

The relation (6.12) is a consequence of statements (a_1) and (a_2), below as we show. For simplicity and without loss of generality we suppose that u_0 is the origin in U.

(a_1) The Lagrange** reduction of a quadratic form yields a nonsingular linear transformation

$$T : \sigma^h = c_{hi} r^i \qquad (h = 1, \ldots, n) \tag{6.13}$$

of R^n into R^n such that under T

---

* The following proof is presented here for the first time.

** See Bôcher [1], pp. 131-136.

$$(6.14) \qquad a_{ij}(u_0) r^i r^j = \|\sigma\|^2 .$$

(a_2) The composition $L \circ T$ of T and of a suitably chosen orthogonal linear transformation L of the space R^n of n-tuples σ onto itself will give a nonsingular linear transformation

$$(6.15) \qquad S : \sigma^h = e_{hi} r^i$$

of R^n onto R^n such that $S(r_0)$ is a nonnull n-tuple $(e, 0, \ldots, 0)$. The transformation S is the contravariant transformation associated with the point transformation

$$(6.16) \qquad H : v^h = e_{hi} u^i$$

of the coordinate domain U onto a coordinate domain $V = H(U)$. Under (6.16) and the associated contravariant transformation S : (6.15) the pair (u_0, r_0) goes into a pair (v_0, σ_0) of which $v_0 = \underset{\sim}{0}$ and $\sigma_0 = (e, 0, \ldots, 0)$.

Proof of (6.12). Under the transformation (6.15)

$$(6.17) \qquad f(u_0, r) = \|\sigma\| \qquad (r \neq \underset{\sim}{0}) .$$

Subject to (6.15)

$$(6.18) \qquad f_{r^i r^j}(u_0, r) = \frac{\partial^2(\|\sigma\|)}{\partial\sigma^h \partial\sigma^k} e_{hi} e_{kj}$$

for i and j on the range $1, \ldots, n$. Since the n-square matrix $\| e_{hi} \|$ is nonsingular, it follows from (6.18) that for $\sigma_0 = (e, 0, \ldots, 0)$

$$(6.19)\qquad \operatorname{rank}\left\| \mathcal{F}_{r^i r^j} \right\|^{(u,r)=(u_0,r_0)} = \operatorname{rank}\left\| \frac{\partial^2(\|\sigma\|)}{\partial\sigma^h \partial\sigma^k} \right\|^{\sigma=\sigma_0} .$$

A simple computation shows that the matrix on the right of (6.19), evaluated when $\sigma_0 = (e, 0, \ldots, 0)$, is a matrix whose nondiagonal elements are zero and whose diagonal reduces to the n-tuple $(0, 1/e, \ldots, 1/e)$. Theorem 6.2 follows.

Hypothesis. *From this point on in this book we shall assume that each W-preintegrand is nonsingular in the sense of Definition* 6.1.

Conditions on preintegrands F. Our major conditions on the preintegrands F depend upon whether we are concerned with the "local" or "global" theory. Theorems, limited to the neighborhood of a pair (u, r) in the domain of F or to the neighborhood of an extremal of J, are regarded as *local*. The existence or properties of an extremal g, joining two points on M_n and of a given homotopy type, are regarded as *global*, as are the relations between the indices of these extremals and the underlying topology. Parts I, II, and III are concerned with the local theory.

For the purposes of the local theory the major conditions on a preintegrand F are that it be *nonsingular* and *positive regular*. A preintegrand F is positive regular if at each 2n-tuple (u, r) in its domain

$$(6.20)\qquad F_{r^i r^j}(u, r)\lambda^i\lambda^j > 0$$

for each n-tuple λ which is contravariant at u and not a scalar multiple of r. See §14 and (4.4).

For the purposes of the global theory the major conditions on the preintegrands F are that they be *nonsingular* and *positive definite*, that is, that $F(u, r) > 0$ for each 2n-tuple (u, r) in the domain of F.

At the end of §18, Theorem 18.2 implies that a preintegrand which is nonsingular and positive-definite is also positive-regular. However, the converse is false, as examples will show.

The R-preintegrands f are nonsingular, as Theorem 6.2 affirms. They are obviously positive definite and hence positive regular as Theorem 18.2 implies. The nonsingularity of the preintegrands F is essential in showing that the Euler equations have solutions. See §7. Parameterized by R-length, these solutions are called <u>extremals</u>.

§7. The Euler-Riemann equations. The difficulties in adequately solving the Euler equations (6.1) under the assumption that F is "nonsingular" in the sense of Def. 6.1 are readily met if one replaces the Euler equations by the set of (n+1) equations

$$\left[\begin{array}{lll} (7.0)' & \frac{d}{dt}F_{r^i}(u,\dot{u}) = F_{u^i}(u,\dot{u}) & (i = 1,\ldots,n) \\ (7.0)'' & \frac{d}{dt}\mathcal{f}(u,\dot{u}) = 0 & \end{array}\right.$$

where F and $\mathcal{f}$ are, respectively, the W- and R-preintegrands associated with a presentation (ϕ, U) in $\mathcal{D}M_n$. Solutions

$$(7.1) \qquad t \to u(t) : [t_0, t_1] \to U$$

of the system (7.0) are required to be regular and hence of at least class C^1. According to Lemma 2.1 the condition (7.0)'', adjoined to the condition (7.0)', is equivalent to the condition that each regular solution of the equations (7.0)' have an RL-parameterization.

We shall call equations (7.0) the Euler-Riemann equations associated with the preintegrands F and $\mathcal{f}$.

We shall give the Euler-Riemann equations (7.0) the following equivalent form

$$\left[\begin{array}{lll} (7.2)' & \frac{du^i}{dt} = r^i & (i = 1,\ldots,n) \\ (7.2)'' & \frac{d}{dt}F_{r^i}(u,r) = F_{u^i}(u,r) & ((u,r) \in U \times \dot{R}^n) \\ (7.2)''' & \frac{d}{dt}\mathcal{f}(u,r) = 0 & \end{array}\right.$$

When the DE (7.2) replace the DE (7.0) the condition that the solutions of the DE : (7.0) be of at least class C^1 and regular, will be replaced by the condition that the domain of the 2n-tuples (u, r) be $U \times \dot{R}^n$ and that solutions of (7.2) be of at least class C^1.

We shall prove the following theorem.

Theorem 7.1. *Let F and $\mathcal{f}$ be W- and R-preintegrands, associated with a presentation (ϕ, U) in $\mathcal{D}M_n$. Let the corresponding associated Euler-Riemann differential equations be given the form* (7.0).

*Corresponding to an arbitrary 2n-tuple** $(u_0, r_0) \in U \times \dot{R}^n$ *and a value* e *in* R, *if* A, B *and* I(e) *are sufficiently small, open connected, neighborhoods of* u_0, r_0, e, *respectively, in* U, $\dot{R}^n$, R, *there exists a* C^∞*-mapping,*

(7.3)' $$(t : a, b) \to \chi(t : a, b) : I(e) \times A \times B \to U$$

such that for fixed n-tuples $a \in A$ *and* $b \in B$ *the partial mapping,*

(7.3)'' $$t \to \chi(t : a, b) : I(e) \to U$$

is a unique $RL^\dagger$*-solution of class* C^∞ *of the Euler-Riemann equations* (7.0) *such that*

(7.4)' $$a = \chi(e : a, b)$$

(7.4)'' $$b = \chi_t(e : a, b) .$$

---

* In Theorem 7.1 the 2n-tuple (u_0, r_0) is fixed and will be termed the 2n-tuple upon which Theorem 7.1 is *based*. This theorem is to be contrasted with Theorem 7.2 which is *based* on an RL-*solution*, given a priori.

† See Lemma 2.2.

<u>A modification</u> (7.6) <u>of the Euler-Riemann equations</u> (7.2). There are two difficulties in solving the Euler-Riemann equations (7.2). The first is that the number $2n$ of unknown variables is one less than the number of DE. This difficulty is met by setting up a system of DE which are a modification of the system (7.2) in that an unknown variable μ is added. With μ added, the unknown variables make up a $(2n+1)$-tuple

(7.5) $$(u^1, \ldots, u^n, r^1, \ldots, r^n, \mu) = (u, r, \mu) .$$

The set of DE

$$\begin{cases} \dfrac{du^i}{dt} = r^i & (i = 1, \ldots, n) \quad (7.6)' \\ \dfrac{d}{dt}\left(F_{r^i}(u, r) + \mu \mathscr{f}_{r^i}(u, r)\right) = F_{u^i}(u, r) + \mu \mathscr{f}_{u^i}(u, r) & \quad (7.6)'' \\ \dfrac{d}{dt}\mathscr{f}(u, r) = 0 & ((u, r, \mu) \in U \times \dot{R}^n \times R) \quad (7.6)''' \end{cases}$$

is well-defined. The unknown $(2n+1)$-tuple, (u, r, μ) is restricted to $U \times \dot{R}^n \times R$.

The following lemma establishes a basic relation between solutions of the Euler-Riemann equations (7.2) and their modification (7.6).

<u>Lemma</u> 7.1 (i). <u>Each</u> C^1<u>-solution</u>

(7.7)' $$t \to (u(t), r(t)) : [t_0, t_1] \to U \times \dot{R}^n$$

<u>of the Euler-Riemann equations</u> (7.2) <u>yields a solution</u>

(7.7)'' $$t \to (u(t)), r(t), \mu(t))$$

<u>of the</u> DE (7.6) <u>in which</u> $\mu(t) \equiv 0$.

(ii) Each* C^2-solution

(7.8) $$t \to (u(t), r(t), \mu(t)) : [t_0, t_1] \to U \times \dot{R}^n \times R$$

of the DE : (7.6) such that $\mu(e) = 0$ for some $e \in [t_0, t_1]$ is such that on $[t_0, t_1]$ $\mu(t) \equiv 0$, and so yields a solution (7.7)' of the Euler-Riemann equations (7.2).

Statement (i) of Lemma 7.1 requires no proof.

To verify (ii) of Lemma 7.1, one multiplies the i-th of the equations (7.6)'' by $\dot{u}^i = r^i$ of the solution (7.8) and sums as to i. On taking account of the identity (4.15), and of the corresponding identity (4.17) when F is replaced by $\mathcal{f}$, one infers from (7.6)'' that

(7.9) $$\dot{u}^i(t)\mathcal{f}_{r^i}(u(t), \dot{u}(t))\dot{\mu}(t) \equiv 0 \qquad (t_0 \leq t \leq t_1) .$$

The homogeneity relation (4.2) enables us to write (7.9) in the form

(7.10) $$\mathcal{f}(u(t), \dot{u}(t))\dot{\mu}(t) \equiv 0 .$$

Since $\mathcal{f}(u, r) \neq 0$ for $(u, r) \in U \times \dot{R}^n$, $\mu(t)$ is a constant. Since $\mu(e) = 0$ by hypothesis, (ii) follows.

A diffeomorphism T. A difficulty in solving the DE (7.6) is that the derivatives as to t of the variables (u, r, μ) are not given as functions of these variables. To meet this difficulty it will be sufficient to restrict the (2n+1)-tuples

---

* Later lemmas will show that a C^1-solution (7.8) of the DE (7.6) is a solution of class C^∞.

(u, r, μ) to so small an open neighborhood Z of $(u_0, r_0, \mu_0)^*$ in R^{2n+1}, that for $(u, r, \mu) \in Z$, the DE (7.6) can be transformed into DE of classical form in $2n+1$ variables

$$(7.11) \qquad (u^1, \ldots, u^n; v^1, \ldots, v^n; w) = (u, v, w) .$$

For i on the range $1, \ldots, n$, a transformation T from a $(2n+1)$-tuple $(u, r, \mu) \in Z$ into a $(2n+1)$-tuple (u, v, w) is defined by setting

$$(7.12) \qquad T : \begin{cases} u^i = u^i & (i = 1, \ldots, n) \\ v^i = F_{r^i}(u, r) + \mu \mathcal{f}_{r^i}(u, r) & \\ w = \mathcal{f}(u, r) & ((u, r, \mu) \in Z) \end{cases}$$

The transformation T satisfies Lemma 7.2.

Lemma 7.2. *If Z is a sufficiently small open neighborhood in $U \times \dot{R}^n \times R$ of the $(2n+1)$-tuple (u_0, r_0, μ_0) the mapping*

$$(7.13) \qquad (u, r, \mu) \to T(u, r, \mu) : Z \to R^{2n+1} ,$$

defined by (7.12), is a diff of class C^∞ of Z into R^{2n+1}, onto an open neighborhood in R^{2n+1} of

$$(7.14) \qquad (u_0, v_0, w_0) = T(u_0, r_0, \mu_0) \qquad (\mu_0 = 0)$$

Since $\mu_0 = 0$, the Jacobian of T at (u_0, r_0, μ_0) is the $(n+1)$-square bordered determinant,

---

* The 2n-tuple (u_0, r_0) is the basic 2n-tuple of Theorem 7.1. We set $\mu_0 = 0$, since solutions of (7.6) on which $\mu(t) \equiv 0$ are sought.

$$(7.15)' \qquad \begin{vmatrix} F_{r^i r^j}(u_0, r_0), & \not{f}_{r^i}(u_0, r_0) \\ \not{f}_{r^j}(u_0, r_0), & 0 \end{vmatrix}$$

Under our hypothesis that F is nonsingular on $U \times \dot{R}^n$, or equivalently, that $F_1(u_0, r_0) \neq 0$, the Jacobian (7.15)' equals

$$(7.15)'' \qquad -F_1(u_0, r_0)\not{f}^2(u_0, r_0) \neq 0 ,$$

in accord with (6.7) and (4.2).

The lemma follows.

<u>Introduction to Lemma</u> 7.3. The inverse of the diff T maps $T(Z)$ onto Z and has a form

$$(7.16) \qquad T^{-1} \begin{cases} u^i = u^i & (i = 1, \ldots, n) \\ r^i = U^i(u, v, w) \\ \mu = M(u, v, w) \end{cases}$$

where the mappings U^i and M are of class C^∞ on $T(Z)$.

For i on the range $1, \ldots, n$ and for $(2n+1)$-tuples $(u, r, \mu) \in Z$ set

$$(7.17) \qquad V^i(u, v, w) = F_{u^i}(u, r) + \mu \not{f}_{u^i}(u, r)$$

subject to (7.16) or equivalently to (7.13). Lemma 7.3 concerns the classical system of DE

$$
\begin{cases}
(7.18)' & \dfrac{du^i}{dt} = U^i(u, v, w) & (i = 1, \ldots, n) \\
(7.18)'' & \dfrac{dv^i}{dt} = V^i(u, v, w) & ((u, v, w) \in T(Z)) \\
(7.18)''' & \dfrac{dw}{dt} = 0 &
\end{cases}
$$

The right members of the DE : (7.18) are of class C^∞. In Lemma 7.3, (7.6)|Z shall denote restrictions of the DE, (7.6) to Z.

Lemma 7.3) (i). Under the diff T, the DE, (7.6)|Z are formally equivalent to the DE (7.18).

(ii). The solutions of the DE (7.18) are mapped biuniquely onto the respective solutions of (7.6)|Z by T^{-1}.

(iii). Since the solutions of the DE (7.18) are of class C^∞, the solutions of (7.6)|Z are likewise of class C^∞.

Statement (i) is a consequence of the definition of V^i in (7.17) and the representation of T in (7.12) and of T^{-1} in (7.16). Statements (ii) and (iii) then follow.

Completion of proof of Theorem 7.1. Classical theorems on DE can be applied to the DE (7.18). One reverses the steps which have led from the solutions of the Euler-Riemann equations (7.0) to the DE (7.18). In order to obtain solutions of the DE (7.6) such that $\mu(t) \equiv 0$, it is sufficient, according to Lemma 7.1 (ii), to obtain solutions of the DE (7.6) for which $\mu(e) = 0$ for some $e \in R$. Let Z^0 be the subset of (2n+1)-tuples (u, r, μ) of Z for which $\mu = 0$. Solutions of (7.18) in T(Z) whose initial values (u(e), v(e), w(e)) are in $T(Z^0)$ will have images under T^{-1} which are solutions of (7.6)|Z for which $\mu(e) = 0$ and hence $\mu(t) \equiv 0$. Dropping $\mu(t)$ from the representation $(u(t), r(t), \mu(t))$ of this subset of solutions of the DE (7.6)|Z one obtains a set of solutions of the DE (7.2) which imply Theorem 7.1.

We shall present a first corollary of Theorem 7.1.

Corollary 7.1. Let (ϕ, U) be a presentation in βM_n and J_F the corresponding W-integral. If a regular arc $\underset{\sim}{u}$ in U is a solution of the Euler equations of J_F, then a reparameterization

$$\underset{\sim}{w} : t \to w(t) : [0, s_1] \to U \tag{7.19}$$

of $\underset{\sim}{u}$ by R-length t, measured from the initial point of $\underset{\sim}{u}$ will be a C^{∞}-solution of the Euler-Riemann equations (7.0).

It follows from Lemma 4.1 that the reparameterization (7.19) of the arc $\underset{\sim}{u}$ must be a solution of the Euler equations (7.0)'. By hypothesis

$$\int_0^t \mathcal{f}(w(a), \dot{w}(a))da = t \qquad (0 \le t \le s_1). \tag{7.20}$$

Hence $\mathcal{f}(w(t), \dot{w}(t)) \equiv 1$ and (7.0)'' of the Euler-Riemann equations is satisfied. It remains to show that the solution $\underset{\sim}{w}$ is of class C^{∞}.

The solution $\underset{\sim}{w}$ has an RL-parameterization. If e is a value in the interval $[0, s_1]$, the 2n-tuple $(w(e), \dot{w}(e))$ is uniquely determined. According to Theorem 7.1 a solution of the Euler-Riemann equations (7.0) with an RL-parameterization is uniquely determined in an open neighborhood I(e) of e by its initial values $(w(e), \dot{w}(e))$ and is of class C^{∞} on I(e).

Thus Corollary 7.1 is true.

Solutions near a prescribed RL-solution. Theorem 7.1 is "based" on a 2n-tuple $(u_0, r_0) \in U \times \dot{R}^n$. Our extension Theorem 7.2 of Theorem 7.1 is based on a mapping

$$(7.21) \qquad t \to z(t) : [t_0, t_1] \to U$$

of class C^∞ which is an RL-parameterized solution of the Euler-Riemann equations (7.0). U is the coordinate domain of a presentation $(\phi, U) \in \mathcal{D}M_n$.

Theorem 7.2. Corresponding to an RL-solution $\underset{\sim}{z}$ (7.21) of the Euler-Riemann equations (7.0) and a value $c \in [t_0, t_1]$, if $\mathcal{A}, \mathcal{B}$ and $I(t_0, t_1)$ are sufficiently small, open, connected neighborhoods of $z(c)$, $\dot{z}(c)$ and $[t_0, t_1]$ respectively in U, $\dot{R}^n$, R, there exists a C^∞-mapping

$$(7.22) \qquad (t : a, b) \to X(t : a, b) : I(t_0, t_1) \times \mathcal{A} \times \mathcal{B} \to U$$

such that for fixed n-tuples $a \in \mathcal{A}$ and $b \in \mathcal{B}$ the partial mapping

$$(7.23) \qquad t \to X(t : a, b) : I(t_0, t_1) \to U$$

is a unique RL-solution of class C^∞ of the Euler-Riemann equations (7.0) such that

$$(7.24)' \qquad a = X(c : a, b)$$

$$(7.24)'' \qquad b = X_t(c : a, b) \quad .$$

Corresponding to a value $t = e$ prescribed in the interval $[t_0, t_1]$ of (7.21) let the values $(z(e), \dot{z}(e))$ on the given solution (7.21) be regarded as a "base"

(u_0, r_0) of a family χ of solutions of the DE (7.0) of the form (7.3)'' for t in an open subinterval $I(e)$ of e in R. Denote this family of solutions by χ^e. In the family χ^e the solution with initial parameters $(a, b) = (u_0, r_0)$ is a subarc of the given solution (7.21) of the DE (7.0) or of an extension* of that solution.

Since the interval $[t_0, t_1]$ is compact, there exists a finite set of values of e, say

$$e_1 < e_2 < \cdots < e_m \tag{7.25}$$

such that $[e_1, e_m] = [t_0, t_1]$ and the corresponding open subintervals of R

$$I(e_1), I(e_2), \ldots, I(e_m) \;, \tag{7.26}$$

defined as was $I(e)$ in the preceding paragraph, are overlapping.† The intervals (7.26) have a union which we denote by $I(t_0, t_1)$. Given the intervals (7.26) let

$$\chi^{e_1}, \chi^{e_2}, \ldots, \chi^{e_m} \tag{7.27}$$

be the corresponding families of RL-solutions of the DE (7.0).

We can suppose that the value $t = c$ given in Theorem 7.2 is one of the values (7.25). Suppose first that $c = t_0$. The family of solutions of the DE (7.0) affirmed to exist in Theorem 7.2, if restricted to the interval $e_1 \leq t \leq e_2$ can be taken as a restriction of the family χ^{e_1}. If the domains $\mathcal{A}$ and $\mathcal{B}$ of the n-tuple

---

*An extension if $e = t_0$ or t_1.

†We can suppose that no one of these intervals overlaps more than one preceding interval.

parameters a and b are sufficiently small open neighborhoods of $u(e_1)$ and $r(e_1)$, the mapping

$$(t, a, b) \to X^{(2)}(t : a, b) : [e_1, e_2] \times \mathcal{A} \times \mathcal{B} \to U ,$$

so defined, can be continued as a C^∞-mapping $X^{(3)}$ of $[e_1, e_3] \times \mathcal{A} \times \mathcal{B} \to U$ by making use of a restriction of the family χ^{e_2} for $e_2 \le t \le e_3$. Continuation of $X^{(3)}$ over the successive intervals

$$[e_3, e_4] \cdots [e_{m-1}, e_m]$$

with the aid of the mappings $\chi^{e_3}, \ldots, \chi^{e_m}$ will be possible if $\mathcal{A}$ and $\mathcal{B}$ are sufficiently small open neighborhoods in R^n of the n-tuples $u(e_1)$ and $r(e_1)$. The C^∞-mapping

$$(t : a, b) \to X^{(m)}(t : a, b) : [e_1, e_m] \times \mathcal{A} \times \mathcal{B} \to U$$

thereby defined can be continued over $I(e_1)$ and $I(e_m)$ to define the required mapping (7.22) backward on $I(e_1)$, forward on $I(e_m)$.

In case $c = t_1$, rather than t_0, the continuation must be in the sense of decreasing t. In case $c = e_i$, where $1 < i < m$, the continuation must be in the sense of increasing t for $e_i < t$ and in the sense of decreasing t for $t < e_i$.

In any case one defines in this way the family X of solutions of (7.0) affirmed to exist in Theorem 7.2.

We shall give two definitions to be extensively used.

Definition 7.1. Extremals and extremaloids of J_F. Let F be a W-preintegrand associated with a presentation $(\phi, U) \in \mathcal{D}M_n$. A regular solution $\underset{\sim}{u}$ of the Euler equations of J_F which is parameterized by R-length, will be called an extremal of J_F. Unless otherwise indicated we suppose that R-length on an extremal is measured from its initial point. A regular solution of the Euler equations of J_F which has an RL-parameterization will be called an extremaloid. An extremal of J_F is an extremaloid.

Definition 7.2. An extremal ξ of* J on M_n. A regular curve ξ on M_n, parameterized by R-length, will be called an extremal of J on M_n if and only if each sufficiently short arc of ξ is representable in the coordinate domain U of some W-preintegrand F by an extremal arc of J_F. Extremaloids of J on M_n are similarly defined.

A regular solution $\underset{\sim}{u}$ of the Euler equations in a coordinate domain U is by definition of class C^1. It may be of no higher class than C^1. However, by Corollary 7.1 a reparameterization of $\underset{\sim}{u}$ by R-length is of class C^∞. Similarly an RL-reparameterization of $\underset{\sim}{u}$ will be of class C^∞.

Definition 7.3. Directions on an extremal in† U_ϕ. Ellipsoids $\mathcal{f}^{u_0}$ of directions at $u_0 \in U$. Let (ϕ, U) be a presentation with F and $\mathcal{f}$ the associated W- and R-preintegrands. If

---

* See §3 for definition of J.

† U_ϕ is the coordinate domain of the presentation (ϕ, U). See §1.

$$(7.28) \qquad \underset{\sim}{u} \to u(s) : [0, s_1] \to U$$

is an extremal of J_F the n-tuple $\dot{u}(s)$ will be called the direction r of $\underset{\sim}{u}$ at the point $u_0 = u(s)$. Since $\underset{\sim}{u}$ is parameterized by R-lengths,

$$(7.29) \qquad 1 = \mathcal{f}(u_0, r) = \left(a_{ij}(u_0) r^i r^j\right)^{\frac{1}{2}} . \qquad \text{(Cf. (2.6).)}$$

Given $u_0 \in U$, the set of n-tuples r such that (7.29) holds, is an ellipsoid in the space R^n, to be denoted by $\mathcal{f}^{u_0}$. It is relative to the presentation (ϕ, U). An n-tuple r satisfying (7.29) is termed R-unitary at u_0.

Essential use of the ellipsoid $\mathcal{f}^{u_0}$ will be made in Sections 9 and 19.

Now that extremals of J on M_n have been defined, a first, but necessarily incomplete view of our global objectives can be given.

Global objectives. Let points A_1 and A_2 be prescribed on M_n, together with a curve η joining A_1 to A_2 on M_n. The class of extremals γ joining A_1 to A_2 of the "homotopy type" of η on M_n is an object of study. By the index of such an extremal γ is meant the count of conjugate* points on γ of A_1 definitely preceding A_2. Let m_k be the number of extremals of J joining A_1 to A_2 of index k and of the "homotopy type" of η. We seek to relate the numbers, $m_0, m_1, m_2, \ldots$ to the homological properties of the Fréchet space $\mathcal{F}_\eta$ of Fréchet classes of curves joining A_1 to A_2 each of the homotopy type of η. See Morse [3]. We shall be concerned with point pairs A_1, A_2 for which each of the numbers m_k is finite. A definition is in order.

---

*Defined in Sections 10 and 23.

Definition 7.4. Curves joining A_1 to A_2 on M_n of the same homotopy type. For a a parameter in the interval $[0,1]$ we shall characterize a continuous family of curves h_a on M_n which join A_1 to A_2 and whose existence implies that the curve h_0 is homotopic to h_1 among curves joining A_1 to A_2.

For each value of $a \in [0,1]$ a curve h_a shall be given by a mapping

$$t \to h(t, a) : [c_a, d_a] \to M_n \qquad (c_a < d_a) \tag{7.30}$$

which is continuous with respect to t. For future applications it is necessary to admit intervals $[c_a, d_a]$ for t whose endpoints vary with a but which are required to vary continuously with a. The curve h_a is to join A_1 to A_2, that is,

$$h(c_a, a) \equiv A_1, \quad h(d_a, a) \equiv A_2 \qquad (0 \leq a \leq 1) . \tag{7.31}$$

Moreover, the mapping (7.30) shall be the "partial mapping" $h(\cdot, a)$ of a continuous mapping,

$$(t, a) \to h(t, a) : X \to M_n \tag{7.32}$$

where

$$X = \{ (t, a) \in R \times [0,1] | c_a \leq t \leq d_a \} .$$

Under the conditions of the preceding paragraph the curve h_0 is said to be A_1A_2-homotopic to the curve h_1.

Definition 7.5. Relative homotopy. Let Z be a subset of curves on M_n joining A_1 to A_2. Two curves η and ζ joining A_1 to A_2 on M_n will be said to be homotopic relative to Z if η, suitably reparameterized,* can be deformed into ζ, suitably reparameterized, through a suitably reparameterized subset of curves of Z.

Definition 7.6. Self-homotopic sets of curves. A set Z of curves joining A_1 to A_2 such that each curve of Z is homotopic to each other curve of Z relative to Z will be called self-homotopic.

Exercise 7.1. Let F and G be preintegrands associated, respectively, with equivalent presentations (ϕ, U) and (ψ, V) in $\mathcal{L}M_n$. Set $\Theta = \phi^{-1} \circ \psi$ and prove the following. A regular mapping

$$\underset{\sim}{v} : t \to v(t) : [t_1, t_2] \to V$$

defines an extremaloid of J_G if and only if the mapping

$$\underset{\sim}{u} : t \to \Theta(v(t)) : [t_1, t_2] \to U$$

defines an extremaloid of J_F. Cf. Corollary 5.1.

Exercise 7.2. When J reduces locally to the integral (2.4)'' of Riemannian length on M_n, extremals are called geodesics on M_n. For $-1 < t < 1$ let $t \to u(t)$ define a local regular representation of a geodesic γ on M_n. Show that the mapping $t \to u(-t)$ represents a geodesic on M_n which traces γ in

---

* Only those parameterizations are admitted which are sense-preserving.

the opposite sense. Show that this statement is false in general when J is not the integral of length on M_n.

§8. From Weierstrass $F(u, r)$ to Euler $f(x, y, p)$. In the study of J_F and, in particular, of conjugate points on extremals, an evaluation of Weierstrass integrals

$$(8.1)\qquad J_F(\underset{\sim}{u}) = \int_{t_0}^{t_1} F(u(t), \dot{u}(t))\, dt$$

as integrals in the Euler nonparametric form

$$(8.2)\qquad J_f(\underset{\sim}{g}) = \int_{x_0}^{x_1} f(x, g(x), g'(x))\, dx$$

is very useful when possible. For piecewise regular curves $\underset{\sim}{u}$ in U along which one of the coordinates $u^1, \ldots, u^n$ can serve as parameter in place of t, such an evaluation is always possible. Which coordinate u^i is preferred is immaterial. We shall prefer u^1.

The following definition is consistent with this preference of u^1.

Definition 8.0. The Euler preintegrand $f(x, y, p)$. With an arbitrary presentation (ϕ, U) and an associated W-preintegrand F, we shall associate an Euler preintegrand f in nonparametric form. Recall that $F(u, r)$ is defined for $(u, r) \in U \times \dot{R}^n$. We set $m = n-1$ and define the values of f by setting

$$(8.3)\qquad f(x, y_1, \ldots, y_m, p_1, \ldots, p_m) = F(x, y_1, \ldots, y_m, 1, p_1, \ldots, p_m)$$

for

$$(8.4)\qquad (x, y_1, \ldots, y_m) \in U,\ (p_1, \ldots, p_m) \in R^m .$$

It is clear that f is of class C^{∞} on its domain. For integers μ, ν on the range $1, \ldots, m = n-1$ and for (x, y, p), as in (8.4),

$$(8.5) \qquad f_{y_\mu} = F_{u^{\mu+1}}; \; f_{p_\mu} = F_{r^{\mu+1}}; \; f_{p_\mu p_\nu} = F_{r^{\mu+1} r^{\nu+1}} .$$

We term the preintegrand f defined by (8.3) the <u>Euler mate</u> of F.

A class of curves which admit reparameterization in U with u^1 as parameter will now be defined.

<u>Definition</u> 8.1. <u>Monge curves on</u> U. A piecewise regular curve

$$(8.6) \qquad \underset{\sim}{M} : t \to M(t) : [t_0, t_1] \to U$$

along which $\dot{M}^1(t) > 0$ (except at corners) will be called a <u>Monge* curve</u> on U. On a Monge curve each coordinate u^i can be represented as a function of class D^1 of the first coordinate u^1.

Along a Monge curve $\underset{\sim}{M}$ of form (8.6) the range of u^1 is the interval $[M^1(t_0), M^1(t_1)]$. An explicit reparameterization of $\underset{\sim}{M}$ will now be defined.

<u>Definition</u> 8.2. <u>The x-parameterized mate of a Monge curve.</u> A reparameterization of the Monge curve $\underset{\sim}{M}$: (8.6) in which u^1 becomes the parameter will be a curve $\underset{\sim}{m}$ in U of the form

$$(8.7) \qquad \underset{\sim}{m} : u^1 \to (m^1(u^1), \ldots, m^n(u^1)) : [M^1(t_0), M^1(t_1)] \to U$$

along which $m^1(u^1) \equiv u^1$.

---

* The curves introduced in Def. 8.1 are called <u>Monge curves</u> because Monge invariably represented curves and surfaces in terms of coordinates as parameters.

Let E_n be the space of rectangular coordinates $(x, y_1, \ldots, y_m)$. Set $[x_0, x_1] = [M^1(t_0), M^1(t_1)]$. By the x-parameterized mate g in E_n of the Monge curve $\underset{\sim}{M}$ of (8.6) is meant a curve in E_n with a graph

$$(8.8) \qquad (y_1, \ldots, y_m) = (g_1(x), \ldots, g_m(x)) \qquad (x_0 \leq x \leq x_1)$$

where

$$(8.9)' \qquad (x, g_1(x), \ldots, g_m(x)) \equiv (m^1(x), \ldots, m^n(x)) \qquad (x_0 \leq x \leq x_1)$$

or equivalently,

$$(8.9)'' \qquad (x, g_1(x), \ldots, g_m(x)) = (M^1(t), \ldots, M^n(t)) \qquad (x_0 \leq x \leq x_1)$$

subject to the condition that $M^1(t) = x$.

The following theorem gives conditions under which a Weierstrass integral, such as (8.1), can be evaluated as an Euler integral, such as (8.2).

Theorem 8.1. *Let F be a W-preintegrand associated with a presentation (ϕ, U) and f the Euler preintegrand defined by (8.3), as the "mate" of F. If $\underset{\sim}{M}$ is a Monge curve of form (8.6) and g its x-parameterized mate, introduced in Def. 8.2, then*

$$(8.10) \qquad \int_{x_0}^{x_1} f(x, g(x), g'(x))\, dx = \int_{t_0}^{t_1} F(M(t), \dot{M}(t))\, dt$$

where $[x_0, x_1] = [M^1(t_0), M^1(t_1)]$.

Because F satisfies the condition (4.1) of homogeneity and the curve $\underset{\sim}{m}$ of (8.7) is a regular reparameterization of the Monge curve $\underset{\sim}{M}$ of (8.6),

(8.11) $$\int_{t_0}^{t_1} F(M(t), \dot{M}(t))\, dt = \int_{x_0}^{x_1} F(m(x), \dot{m}(x))\, dx \; .$$

On taking account of the fact that for $x_0 \leq x \leq x_1$, $m^1(x) \equiv x$ and hence $\dot{m}^1(x) \equiv 1$, and of the definition of f in (8.3) and of g in (8.9), one infers that

(8.12) $$\int_{x_0}^{x_1} F(m(x), \dot{m}(x))\, dx = \int_{x_0}^{x_1} f(x, g(x), g'(x))\, dx \; .$$

The relation (8.10) follows from (8.11) and (8.12).

The following lemma is a consequence of the assumption that the Weierstrass coefficient $F_1(u, r) \neq 0$ for each $2n$-tuple (u, r) in the domain $U \times \dot{R}^n$ of F.

Lemma 8.1. *The Euler preintegrand f which is the mate of the W-preintegrand F is such that for μ, ν on the range, $1, \ldots, m = n-1$, the m-square determinant,*

(8.13) $$\Delta(x, y, p) = |f_{p_\mu p_\nu}(x, y, p)| \neq 0$$

for each $(2m+1)$-tuple (x, y, p) in the domain of f.

For (x, y, p) in the domain of f, set

(8.14) $$(\hat{u}, \hat{r}) = (x, y_1, \ldots, y_m; 1, p_1, \ldots, p_m) \; .$$

Let $A(\hat{u}, \hat{r})$ be the cofactor of the element $F_{r^1 r^1}(\hat{u}, \hat{r})$ in the determinant $|F_{r^i r^j}(\hat{u}, \hat{r})|$. It follows from the third set of relations in (8.5) that $A(\hat{u}, \hat{r}) = \Delta(x, y, p)$ when (8.14) holds. From (6.9) we infer that

$$A(\hat{u}, \hat{r}) = F_1(\hat{u}, \hat{r})\hat{r}^1\hat{r}^1 . \tag{8.15}$$

Now $\hat{r}^1 = 1$ by virtue of (8.14), so that $A(\hat{u}, \hat{r}) \neq 0$. Since $A(\hat{u}, \hat{r}) = \Delta(x, y, p)$, the proof of Lemma 8.1 is complete.

The following lemma has important applications.

<u>Lemma</u> 8.2. *A regular Monge curve in* U *of form* $\underset{\sim}{M}$: (8.6) *is a solution of the* n *Euler equations of* J_F, *if and only if its* x-*parameterized mate* g *of Def.* 8.2 *is a solution of the* n-1 *Euler equations,*

$$\frac{d}{dx} f_{p_\mu}(x, y, y') - f_{y_\mu}(x, y, y') = 0 , \qquad (\mu = 1, \ldots, n-1) \tag{8.16}$$

where f *is the Euler mate of* F.

If $\underset{\sim}{M}$ is a C^1-solution of the Euler equations of J_F, its u^1-reparameterization $\underset{\sim}{m}$ of (8.7) is also a solution of the Euler equations of J_F in accord with Lemma 4.1. With u^1 replaced by x, as parameter, Lemma 4.1 implies that for $i = 1, \ldots, n$,

$$\frac{d}{dx} F_{r^i}(m(x), \dot{m}(x)) - F_{u^i}(m(x), \dot{m}(x)) \equiv 0 \qquad (x_0 \leq x \leq x_1) . \tag{8.17}$$

Now $m^1(x) \equiv x$ and $\dot{m}^1(x) \equiv 1$, so that in accord with (8.9), (8.17) can be written in the form

(8.18) $$\frac{d}{dx} F^x_{r^i} - F^x_{u^i} \equiv 0 \qquad (i = 1, \ldots, n)$$

where the superscript x means evaluation for

(8.19) $$(u, r) = (x, g_1(x), \ldots, g_m(x); 1, g_1'(x), \ldots, g_m'(x)) .$$

From equations (8.18) with $i = 2, \ldots, n$ and from the relations (8.5), we infer that g satisfies the n-1 Euler equations (8.16).

It remains to prove the following:

(a) If the mate g of a Monge curve $\underset{\sim}{M}$ of form (8.6) is a solution of the n-1 Euler equations (8.16), then $\underset{\sim}{M}$ satisfies the n Euler equations of J_F.

Proof of (a). Since the m-square determinant (8.13) is not zero, the solution $x \to g(x)$ of the Euler equations (8.16) must be of class C^2 at least. By hypothesis g is the mate of $\underset{\sim}{M}$. Hence there exists a reparameterization $\underset{\sim}{m}$ of $\underset{\sim}{M}$ such that (8.9)' holds. For $\underset{\sim}{m}$, identities of the form (8.17) hold for $i = 2, \ldots, n$ because g satisfies the n-1 Euler equations (8.16), $m^1(x) \equiv x$, and (8.5) holds.

That (8.17) holds when $i = 1$, follows from (4.15),* since $\dot{m}^1(x) \equiv 1$ for $x_0 \le x \le x_1$. Finally $\underset{\sim}{M}$, like $\underset{\sim}{m}$, satisfies the Euler equations of J_F, in accord with Lemma 4.1.

Thus Lemma 8.2 is true.

---

* Lemma 4.2 applies since the mapping $x \to m(x)$ is of class C^2 at least.

Simple and nonsimple extremals of J on M_n. Our definition of conjugate points on an extremal of J in Chapter 4 and our study of minimizing extremals in Chapters 5 and 6 will be restricted to extremals on M_n which are simple. We shall see in Chapter 8 how theorems on nonsimple extremals, which are essential for our purposes, can be inferred from the corresponding theorems on simple extremals.

Chapter 4

Conjugate points

§9. Proper polar families of extremal arcs. To define conjugate points on extremal arcs of the W-integral J on M_n it will be sufficient to define conjugate points on extremals of a W-integral J_F in a coordinate domain U. This is because each curve on M_n which is both regular and simple admits a representation in the coordinate domain U of a presentation* in $\mathcal{D}M_n$. Extremals of J which are not simple will be treated in Sections 22 and 23 by an extension of the methods of this section.

Let (ϕ, U) be a presentation in $\mathcal{D}M_n$ and J_F the associated W-integral on U. To define conjugate points on extremal arcs of J_F in U, we shall define "proper polar families" $\underset{\sim}{\Phi}$ of extremal arcs of J_F in U issuing from a point of U. Use will be made of these families to define conjugate points on an extremal

$$(9.1) \qquad \underset{\sim}{z} : s \to z(s) : [0, s_1] \to U,$$

called the central extremal of the family.

Each extremal of the family is supposed parameterized by R-length s, measured from its initial point over an interval $[0, s_1]$. A parameter α identifying an extremal Γ_α of the family, will be an m-tuple α, restricted to an open origin-centered m-ball B_m in R^m. The family† $\underset{\sim}{\Phi}$ will be defined by a mapping

---

* See Theorem 22.1.

† We denote the family by $\underset{\sim}{\Phi}$ and the mapping (9.2)' defining the family by Φ.

(9.2)' $$(s, a) \to \Phi(s, a) : [0, s_1] \times B_m \to U \qquad (m = n - 1)$$

of class C^∞, such that for each m-tuple $a \in B_m$ the partial mapping

(9.2)'' $$\Phi(\cdot, a) : s \to \Phi(s, a) : [0, s_1] \to U$$

is an extremal Γ_a of the family.

To be useful in defining conjugate points the family $\underset{\sim}{\Phi}$ should be both polar and proper, in senses which will now be defined.

Polar families. The above family $\underset{\sim}{\Phi}$ of extremal arcs of J_F will be *said to have a* pole $u_0 \in U$, *if for some value* $c \in [0, s_1]$

(9.3) $$\Phi(c, a) \equiv u_0 \qquad (a \in B_m) .$$

Proper polar families. A pole u_0 at $s = c$ of an extremal family $\underset{\sim}{\Phi}$ will be termed proper, if Φ_s has a functional matrix* such that

(9.4) $$\text{rank} \left\| \frac{\partial \Phi_s^i(s, a)}{\partial a_\mu} \right\| = n-1 \qquad \begin{matrix} (i = 1, \ldots, n) \\ (\mu = 1, \ldots, m) \end{matrix}$$

when $s = c$ and $a = \underset{\sim}{0}$. A family $\underset{\sim}{\Phi}$ with a proper pole will be termed a proper polar family.

The Jacobian $D_\Phi^c(s)$ and matrix $\underset{\sim}{D}_\Phi^c(s)$. The extremal $\Gamma_0 = \Phi(\cdot, \underset{\sim}{0})$, will be called the central extremal of the family $\underset{\sim}{\Phi}$. We introduce the Jacobian

---

* The matrix (9.4) has n rows indexed by i and n-1 columns indexed by μ. Each column has n elements.

$$(9.5) \qquad D^c_\Phi(s) = \frac{D(\Phi^1, \Phi^2, \ldots, \Phi^n)}{D(s, a_1, \ldots, a_m)}(s,a)\Big|^{a=\underset{\sim}{0}} .$$

The n-square matrix whose elements are the entries of $D^c_\Phi(s)$ will be denoted by $\underset{\sim}{D}{}^c_\Phi(s)$. We shall refer to the nullity of $\underset{\sim}{D}{}^c_\Phi(s)$.

Objective of §9. Four lemmas will be proved, each of which concerns a determinant $D^c_\Phi(s)$ and the corresponding matrix $\underset{\sim}{D}{}^c_\Phi(s)$. These lemmas will justify our use in §10 of the zeros of $D^c_\Phi(s)$, other than $s = c$, to define* the conjugate points of $s = c$ on the central extremal Γ_0 of the family $\underset{\sim}{\Phi}$ and the use of the nullities of the corresponding matrix $\underset{\sim}{D}{}^c_\Phi(s)$ to define the multiplicities of these conjugate points.

A first lemma follows.

Lemma 9.1. *Let $\underset{\sim}{\Phi}$ be a family of extremal arcs of the integral J_F, with proper pole u_0 at $s = c$. The Jacobian*† *$D^c_\Phi(s)$ of the mapping Φ is such that*

$$(9.6) \qquad D^c_\Phi(s) = (s-c)^m N^c(s) , \qquad (0 \leq s \leq s_1)$$

where the mapping $s \to N^c(s)$ is of class C^∞ for $(0 \leq s \leq s_1)$ and $N^c(c) \neq 0$.

Set $\Phi_s(c, a) = r(a)$ for $a \in B_m$. The n-tuple $r(a)$ is a point r on the ellipsoid

$$(9.7) \qquad \mathcal{J}^{u_0} : \mathcal{J}(u_0, r) = (a_{ij}(u_0) r^i r^j)^{\frac{1}{2}} = 1 \qquad \text{(of Def. 7.3)} .$$

---

* In the classical nonparametric theory A. Kneser has made effective use of Jacobians of extremal families to represent conjugate points. See Morse [2], p. 26.

† The Jacobian of the mapping Φ is evaluated when $a = \underset{\sim}{0}$.

The n-tuple $r(\underset{\sim}{0})$ gives the first column of $\underset{\sim}{D}^c_\Phi(c)$. Since (9.3) holds, each of the last m columns of $\underset{\sim}{D}^c_\Phi(s)$ vanishes when $s = c$. It follows that

$$(9.8) \qquad D^c_\Phi(s) = (s-c)^m N^c(s), \qquad (0 \leq s \leq s_1)$$

where the mapping $s \to N^c(s)$ is of class C^∞. Cf. Morse [1], p. 24.

Proof that $N^c(c) \neq 0$. $N^c(c)$ is given by the n-square determinant,

$$(9.9) \qquad |r(\underset{\sim}{0}), r_{a_1}(\underset{\sim}{0}), \ldots, r_{a_m}(\underset{\sim}{0})| .$$

Since the n-tuple $r(a)$ is a point r on the ellipsoid $\mathscr{E}^{u_0}$, (9.7), for each $a \in B_m$, the m vectors $r_{a_\mu}(\underset{\sim}{0})$, $\mu = 1, \ldots, m$, have directions tangent in R^m to the ellipsoid $\mathscr{E}^{u_0}$ at the point $r(\underset{\sim}{0})$. These m-tuples $r_{a_\mu}(\underset{\sim}{0})$ are columns of the matrix (9.4) and so are linearly independent by hypothesis. Since the vector $r(\underset{\sim}{0})$ is not tangent to the ellipsoid $\mathscr{E}^{u_0}$ at the point $r(\underset{\sim}{0})$, we infer that $N^c(c)$, as given by the determinant (9.9), fails to vanish.

We continue with the presentation $(\phi, U) \in \mathcal{D}M_n$ and associated W-preintegrand F and state an existence lemma.

Lemma 9.2. Given an extremal arc of J_F of form (9.1), or explicitly,

$$(9.10) \qquad \underset{\sim}{z} : s \to z(s) : [0, s_1] \to U$$

and a value $c \in [0, s_1]$, there exists a proper polar family $\underset{\sim}{\Phi}$ of extremals of J_F with pole $z(c)$ and with $\underset{\sim}{z}$ of (9.10) as its central extremal.

Theorem 7.2 is "based" on the RL-parameterized solution (7.21) of (7.0). The solution (7.21) of (7.0) can be identified with the extremal $\underset{\sim}{z}$ of

(9.10), provided $[t_0, t_1] = [0, s_1]$. Theorem 7.2 then implies the existence of a family of extremaloids (Def. 7.1)

$$t \to X(t : a, b) : I(0, s_1) \to U \tag{9.11}$$

which contains the extremal arc

$$t \to X(t : z(c), \dot{z}(c)) : I(0, s_1) \to U . \tag{9.12}$$

This extremal is an open extension of the extremal arc $\underset{\sim}{z}$ of Lemma 9.2 on which $0 \leq s \leq s_1$.

<u>Definition of</u> $\underset{\sim}{\Phi}$. The family of extremals of J_F affirmed to exist in Lemma 9.2, can be obtained by suitably restricting the family of extremaloids given by (9.11).

To that end the mappings (9.11) are first restricted to the interval $[0, s_1]$. The n-tuple a in (9.11) is then fixed as the point $u_0 = z(c)$. The resultant extremaloids (9.11) meet the point u_0 when $t = c$. In order to cut this family down to extremaloids which are extremals, the n-tuples b in (9.11) must be restricted to points r on the ellipsoid,

$$\mathcal{f}(u_0, r) = (a_{ij}(u_0)r^i r^j)^{\frac{1}{2}} = 1. \qquad \text{(Cf. (7.29).)} \tag{9.13}$$

One has the relation $\mathcal{f}(z(c), \dot{z}(c)) = 1$, since $\underset{\sim}{z}$ is an extremal of J_F. One seeks extremals of J_F which meet $u_0 = z(c)$ with "directions" at u_0 near the direction of $\underset{\sim}{z}$ at $z(c)$. Equivalently in (9.13) one seeks points $r \in R^n$ which are near $r_0 = \dot{z}(c)$ on the ellipsoid $\mathcal{f}^{u_0}$.

Since the ellipsoid is a regular real analytic m-manifold in R^n the following is true. If B_m is an origin-centered open m-ball in R^m, there exists a C^∞-diff

$$(9.14) \qquad \alpha \to \pi(\alpha) : B_m \to \mathcal{J}^{u_0}$$

onto an open neighborhood N of the point $r_0 = \dot{z}(c)$ on $\mathcal{J}^{u_0}$, such that $N \subset \mathcal{B}$ of Theorem 7.2. We can then define Φ by setting

$$(9.15) \qquad X(s : z(c), \pi(\alpha)) = \Phi(s, \alpha)$$

for $\alpha \in B_m$ and $0 \leq s \leq s_1$. It remains to prove the following:

(i) <u>The condition</u> (9.4) <u>is satisfied by</u> Φ. To verify (i), note first that

$$(9.16) \qquad \Phi_s(c,\alpha) = X_s(c : z(c), \pi(\alpha)) \equiv \pi(\alpha) \qquad (\alpha \in B_m)$$

in accord with (9.15) and (7.24)''. Moreover, the functional matrix of π, evaluated at $\alpha = \underset{\sim}{0}$, is of rank m, since (9.14) is a diff. Thus the condition (9.4) on Φ is satisfied.

This completes the proof of Lemma 9.2.

The conjugate points of a point on an extremal should be shown to be independent of the particular polar family of extremals used to define these conjugate points. The following lemma serves this purpose.

<u>Lemma</u> 9.3. <u>Let</u> $\underset{\sim}{\Phi}$ <u>and</u> $\underset{\sim}{\Phi}'$ <u>be two proper polar families of extremal arcs of</u> J_F <u>in</u> U <u>of general form</u> (9.2)' <u>with the same central extremal</u> $\underset{\sim}{z}$

and the same pole $u_0 = z(c)$. *There then exists a nonsingular n-square matrix* H *of real constants such that the matrix product*

$$(9.17) \qquad \underset{\sim}{D}^{c}_{\Phi'}(s) \cdot H \equiv \underset{\sim}{D}^{c}_{\Phi''}(s) \qquad (0 \leq s \leq s_1) .$$

The proof of Lemma 9.3 is begun by setting

$$(9.18)' \qquad \Phi'_s(c, a) = \pi'(a), \quad \Phi''_s(c, a) = \pi''(a) \qquad (a \in B^{\rho})$$

where B^{ρ} is an origin-centered open m-ball of radius ρ in the origin-centered open m-ball B_m of (9.2)'. If ρ is sufficiently small, Theorem 7.2 implies that, when (9.18)' holds,

$$(9.18)'' \qquad \Phi'(s,a) \equiv X\big(s : z(c), \pi'(a)\big); \quad \Phi''(s,a) \equiv X\big(s : z(c), \pi''(a)\big)$$

for $0 \leq s \leq s_1$ and $a \in B^{\rho}$. The condition (9.4) on Φ'_s and Φ''_s implies that the functional matrices of $\pi'(a)$ and $\pi''(a)$, if evaluated when $a = \underset{\sim}{0}$, have the rank $m = n-1$.

If N is a sufficiently small open neighborhood of the origin in R^{n-1}, the restrictions $\pi'|N$ and $\pi''|N$ are diffs of N onto open neighborhoods of the point $r_0 = \dot{z}(c)$ on the ellipsoid $\mathcal{E}^{u_0}$ of (9.7), where $u_0 = z(c)$. If N_1 is a sufficiently small open neighborhood of the origin in N, there exists a C^{∞}-diff $a \to \kappa(a)$ of N_1 onto an open neighborhood of the origin in N, leaving the origin fixed and such that $\pi''(a) \equiv \pi'(\kappa(a))$ for $a \in N_1$. It follows from (9.18)'' that

(9.19) $$\Phi''(s,\alpha) \equiv \Phi'(s,\kappa(\alpha)) \qquad (0 \leq s \leq s_1, \ \alpha \in N_1) .$$

Let H be the n-square functional matrix of the mapping $[s, \alpha] \to [s, \kappa(\alpha)]$, evaluated when $[s, \alpha] = (s, \underset{\sim}{0})$. The matrix relation (9.17) now follows from the fact that $\underset{\sim}{z}$ is the central extremal of both $\underset{\sim}{\Phi}'$ and $\underset{\sim}{\Phi}''$, and from the identity (9.19).

The following corollary of Lemma 9.3 refers to nullities that will be interpreted in §10 as multiplicities of conjugate points.

Corollary 9.1. Under the hypothesis of Lemma 9.3

(9.20) $$\text{nullity } \underset{\sim}{D}^c_{\Phi'}(s) = \text{nullity } \underset{\sim}{D}^c_{\Phi''}(s) \qquad (0 \leq s \leq s_1) .$$

The matrices on the right and left of (9.17) are identical. By a law of Sylvester,* the rank of the matrix on the right of (9.17) is the rank of $\underset{\sim}{D}^c_{\Phi'}(s)$, since H is nonsingular. Hence (9.20) holds.

Introduction to Lemma 9.4. The conjugate points of a point on an extremal should be shown to be independent of the particular coordinate domain in which they are defined. Lemma 9.4 serves this purpose.

Lemma 9.4 presupposes equivalent presentations (ϕ, U) and (ψ, V) with associated W-integrals J_F and J_G. It shows how a proper polar family $\underset{\sim}{\Psi}$ of extremals of J_G in V leads to an essentially equivalent proper polar family $\underset{\sim}{\Phi}$ of extremals of J_F in U. This lemma is essential in showing

* Cf. Bôcher [1] Theorem 7, page 79.

that the properties* of conjugate points, already established in the non-parametric theory, can be carried over to the parametric theory.

A proper family $\underset{\sim}{\Psi}$ of extremal arcs

$$s \to \Psi(s, \alpha) : [0, s_1] \to V \qquad (\alpha \in B_m) \tag{9.21}$$

of J_G with a pole $v_0 \in V$, when $s = c$, is defined as was a proper family $\underset{\sim}{\Phi}$ of extremal arcs of J_F on U with pole $u = u_0$ when $s = c$. For i on the range $1, \ldots, n$ and μ on the range $1, \ldots, m = n-1$ the $(n \times m)$-matrix

$$\left\| \frac{\partial \Psi^i_s}{\partial \alpha_\mu}(s, \alpha) \right\|^{(c, \underset{\sim}{0})}, \qquad (0 \leq s \leq s_1) \tag{9.22}$$

by analogy with (9.4), is required to have a rank $n-1$. When $\alpha = \underset{\sim}{0}$, (9.21) gives the central extremal, $\underset{\sim}{w} : s \to w(s)$ of the family $\underset{\sim}{\Psi}$. The Jacobian

$$D^c_\Psi(s) = \frac{D(\Psi^1, \Psi^2, \ldots, \Psi^n)}{D(s, \alpha_1, \ldots, \alpha_m)}(s, \alpha)\Bigg|^{\alpha = 0} \qquad (0 \leq s \leq s_1) \tag{9.23}$$

is defined along the "central" extremal $\underset{\sim}{w}$ of the family $\underset{\sim}{\Psi}$ and the corresponding functional matrix denoted by $\underset{\sim}{D}^c_\Psi(s)$.

<u>Lemma</u> 9.4. <u>Let</u> (ϕ, U) <u>and</u> (ψ, V) <u>be equivalent presentations and</u> $\Theta = \phi^{-1} \circ \psi$ <u>the transition diff of</u> V <u>onto</u> U. <u>Let</u> $\underset{\sim}{\Psi}$ <u>be a proper family of extremal arcs of</u> J_G <u>on</u> V <u>with a pole</u> $v_0 \in V$ <u>when</u> $s = c$. <u>If one sets</u>

$$\Phi(s, \alpha) = \Theta(\Psi(s, \alpha)) \qquad (0 \leq s \leq s_1, \ \alpha \in B_m) \tag{9.24}$$

---
* Such as the generalized Sturm Separation Theorem. Cf. Morse [1], p. 141 and Theorem 10.4 of this book.

<u>the resultant mapping,</u>

(9.25) $$(s, a) \to \Phi(s, a) : [0, s_1] \times B_m \to U$$

<u>defines a proper polar family of extremal arcs of</u> J_F <u>on</u> U <u>with a central extremal</u> $\underset{\sim}{z} = \Theta \cdot \underset{\sim}{w}$ <u>and pole</u> $u_0 = \Theta(v_0)$. <u>It follows that</u>

(9.26) $$D^c_\Phi(s) \equiv \frac{D(\Theta^1, \ldots, \Theta^n)}{D(v^1, \ldots, v^n)} (v)\, D^c_\Psi(s) \qquad (0 \le s \le s_1)$$

<u>with the Jacobian of</u> Θ <u>evaluated for</u> $v = w(s)$. <u>Moreover</u>

(9.27) $$\text{nullity } \underset{\sim}{D}{}^c_\Phi(s) = \text{nullity } \underset{\sim}{D}{}^c_\Psi(s) \qquad (0 \le s \le s_1) .$$

The pole u_0 of the family Φ is proper if the $(n \times m)$-matrix

(9.28) $$\left\| \frac{\partial \Phi^i_s}{\partial a_\mu}(s, a) \right\|^{(c, 0)}$$

has the rank $m = n-1$. That this rank is m is seen as follows.

Relative to the transition diff Θ of V onto U, each of the m columns of the matrix (9.28) is the contravariant image at $u_0 = \Theta(v_0)$ of the corresponding column at v_0 of the matrix (9.22). No nontrivial linear combination of the columns of the matrix (9.28) is null, since the same linear combination of the corresponding columns of the matrix (9.22) is nonnull by hypothesis. Hence u_0 is a <u>proper</u> pole of the family $\underset{\sim}{\Phi}$.

The identity (9.26) follows from (9.24). To establish (9.27), let $\underset{\sim}{D}_\Theta(s)$ be the functional matrix of the transition diff Θ, evaluated when the n-tuple

$v = w(s)$. For $0 \leq s \leq s_1$

(9.29)
$$\underset{\sim}{D}_{\Phi}^{c}(s) = \underset{\sim}{D}_{\Theta}(s) \underset{\sim}{D}_{\Psi}^{c}(s) \ .$$

Since the matrix $\underset{\sim}{D}_{\Theta}(s)$ is nonsingular, (9.27) follows from (9.29) and the law of Sylvester used in proving Corollary 9.1.

§10. Conjugate points and their multiplicities. Let (ϕ, U) be a presentation in $\mathcal{L}M_n$ and F the associated W-preintegrand. Let

(10.1) $$\underset{\sim}{z} : s \to z(s) : [0, s_1] \to U$$

be an extremal arc* of J_F. The definition for which the lemmas of §9 have prepared the way can now be given.

Definition 10.0 (i). The conjugate points on the extremal $\underset{\sim}{z}$: (10.1) of J_F of a point† $s = c$ on $\underset{\sim}{z}$ are the points s on $\underset{\sim}{z}$, other than $s = c$, at which $D^c_\Phi(s) = 0$, where $\underset{\sim}{\Phi}$ is any proper polar extremal family in U, with $\underset{\sim}{z}$ as central extremal and $z(c)$ as pole.

(ii) The multiplicity of a conjugate point s of the point $s = c$ on $\underset{\sim}{z}$ is the nullity of the matrix $\underset{\sim}{D}^c_\Phi(s)$.

It follows from the matrix identity (9.17) of Lemma 9.3, that the conjugate points of a point $s = c$ on an extremal $\underset{\sim}{z}$ of J_F, together with their multiplicities are independent of the extremal family $\underset{\sim}{\Phi}$ with pole $z(c)$ and central extremal $\underset{\sim}{z}$ in terms of which the conjugate points of $s = c$ on $\underset{\sim}{z}$ and their multiplicities are defined.

The above definition is "generic." That is, it is supposed made in the same way when the following three replacements are made:

(1) (ϕ, U) by a presentation (ψ, V) as in Lemma 9.4;

(2) the family $\underset{\sim}{\Phi}$ of extremals of J_F in U by the family $\underset{\sim}{\Psi}$ of extremals of J_G in V;

---

*Recall that a curve is called an arc if and only if it is simple.

†It should be understood that a conjugate point of $s = c$ on $\underset{\sim}{z}$ may precede or follow $s = c$.

(3) the central extremal $\underset{\sim}{z}$ of $\underset{\sim}{\Phi}$ by the central extremal $\underset{\sim}{w}$ of $\underset{\sim}{\Psi}$.

The following theorem is a consequence of the lemmas of §9. The proof is left to the reader.

Theorem 10.1. *Let* (ϕ, U) *and* (ψ, V) *be equivalent presentations and* $\Theta = \phi^{-1} \circ \psi$ *the transition diff of* V *onto* U. *Let*

$$\underset{\sim}{z} : s \to \underset{\sim}{z}(s) : [0, s_1] \to U \tag{10.2)'}$$

$$\underset{\sim}{w} : s \to \underset{\sim}{w}(s) : [0, s_1] \to V \tag{10.2)''}$$

be extremal arcs respectively of J_F *and* J_G *such that** $\Theta \cdot \underset{\sim}{w} = \underset{\sim}{z}$. *Then* $\underset{\sim}{w}$ *and* $\underset{\sim}{z}$ *are conjugatewise equivalent in that a point* $s \neq c$ *is conjugate on* $\underset{\sim}{w}$ *to a point* $s = c$ *on* $\underset{\sim}{w}$ *and has the multiplicity* μ *if and only if the point* s *is conjugate on* $\underset{\sim}{z}$ *to the point* $s = c$ *on* $\underset{\sim}{z}$ *and has the multiplicity* μ *on* $\underset{\sim}{z}$.

Properties of conjugate points have been given an extensive development in Morse [1] in the case in which the integral has the classic Euler nonparametric form. Cf. §8. To make use of the theorems on conjugate points when the integral has the Euler form, it is sufficient to apply Theorem 10.1 in the case when one of the two extremal arcs (10.2)' and (10.2)" has a "Monge nonparametric form." For this purpose §22 on "Tubular mappings" has been written. We shall make use of Theorem 10.1 but only after one of the extremals, say $\underset{\sim}{w}$, has been given in a coordinate domain V and the other, $\underset{\sim}{z}$, replaced by an extremal $\bar{z}$ which is an arc of the u^1-axis in a coordinate domain U.

---

* See footnote to Lemma 3.1.

To apply Theorem 10.1, U and V should be domains of equivalent presentations.

Reduction to nonparametric form. If the extremal arc $\underset{\sim}{w}$ of Theorem 10.1 is given and if V is a sufficiently small open neighborhood of the carrier of $\underset{\sim}{w}$, Theorem 22.1 implies the following. There exists a presentation (ϕ, U) equivalent to the presentation (ψ, V) as in Theorem 10.1, and such that the extremal arc $\bar{\underset{\sim}{z}} = \Theta \cdot \underset{\sim}{w}$ in U has for carrier the segment of the u^1-axis in U on which $0 \leq u^1 \leq s_1$ and along which

$$\bar{z}^1(s) \equiv s \equiv u^1 \qquad (0 \leq s \leq s_1) . \tag{10.3}$$

As in Theorem 10.1, let $F(u, r)$ be the preintegrand associated with the presentation (ϕ, U). J_F has the extremal $\bar{\underset{\sim}{z}} = \Theta \cdot \underset{\sim}{w}$.

Definition 10.1. The axial extremal* $\bar{\underset{\sim}{z}}$. We shall term the above extremal arc $\Theta \cdot \underset{\sim}{w}$ an axial extremal, equivalent in U to the extremal $\underset{\sim}{w}$ in V. Each coordinate of $\bar{\underset{\sim}{z}}$ in R^n vanishes identically except its first coordinate; for this coordinate (10.3) holds.

Definition 10.2. The x-parameterized mate g of $\bar{\underset{\sim}{z}}$. As in §8 we refer to the Euclidean space E_n of rectangular coordinates $x, y_1, \ldots, y_m$, where $m = n-1$. In the notation of §8 the x-parameterized "mate" of the axial extremal $\bar{\underset{\sim}{z}}$ in U has the graph

$$g : y_\mu = g_\mu(x) \qquad (\mu = 1, 2, \ldots, m = n-1) \tag{10.4$'$}$$

---

* The bar above $\underset{\sim}{z}$ in $\bar{\underset{\sim}{z}}$ is meant to indicate that $\bar{\underset{\sim}{z}}$ is an axial extremal.

where

$$(10.4)'' \qquad (x, g_1(x), \ldots, g_m(x)) \equiv (\bar{z}_1(x), \ldots, \bar{z}_n(x)) \qquad (0 \leq x \leq s_1) .$$

Since $\bar{\underset{\sim}{z}}$ is axial, $\bar{z}_1(x) \equiv x$ and

$$(10.4)''' \qquad g_1(x) \equiv \cdots \equiv g_m(x) \equiv 0 \qquad (0 \leq x \leq s_1) .$$

Let f be the Euler mate of the W-preintegrand F, defined as in (8.3). The values $f(x, y, p)$ of f are defined for $x, y_1, \ldots, y_m$ an arbitrary n-tuple in U and $(p_1, \ldots, p_m)$ an arbitrary m-tuple in R^m. According to Lemma 8.2 the mapping $x \to g_\mu(x)$, $\mu = 1, \ldots, m$, is a solution of the Euler equations

$$(10.5) \qquad \frac{d}{dx} f_{p_\mu}(x, y, y') = f_{y_\mu}(x, y, y') \qquad (\mu = 1, \ldots, m) .$$

As affirmed by Lemma 8.1, the Euler equations (10.5) are nonsingular in that the m-square determinant

$$(10.6) \qquad |f_{p_\mu p_\nu}(x, y, p)| \neq 0$$

for (x, y, p) in the domain of f.

<u>A proper polar family</u> $\underset{\sim}{\bar{\Phi}}$ <u>with axial central extremal</u> $\bar{\underset{\sim}{z}}$. Such a family exists, by virtue of Lemma 9.2, with a point $\bar{z}(c)$ of $\bar{\underset{\sim}{z}}$ prescribed as a proper pole. Such a family will be termed <u>axial</u> because it has the axial extremal $\bar{\underset{\sim}{z}}$ as central extremal.

Extremals of $\underset{\sim}{\overline{\Phi}}$ x-parameterized. If β_m is a sufficiently small origin-centered m-ball in R^m, then for $a \in \beta_m$ each extremal Γ_a of $\underset{\sim}{\overline{\Phi}}$ is a Monge curve in U (Def. 8.1) and accordingly has an x-parameterized mate, say g^a, which is an extremal of J_f (Lemma 8.2). As seen in §8, g^a is obtained by giving the extremal

$$(10.7)' \qquad s \to \overline{\Phi}(s, a) : [0, s_1] \to U \qquad (a \in \beta_m)$$

an x-parameterization

$$(10.7)'' \qquad x \to Y(x, a) : K_a \to R^m$$

subject to the condition

$$(10.8) \qquad \overline{\Phi}^1(s, a) = x \qquad (0 \leq s \leq s_1;\ a \in \beta_m),$$

where the interval K_a for x on g^a has the form,

$$(10.9) \qquad K_a = [\overline{\Phi}^1(0, a), \overline{\Phi}^1(s_1, a)] \qquad (a \in \beta_m).$$

As indicated in §8, Y(x, a) is defined by setting

$$(10.10) \qquad (x, Y_1(x, a), \ldots, Y_m(x, a)) = (\overline{\Phi}^1(s, a), \ldots, \overline{\Phi}^n(s, a)) \qquad (\text{cf. } (8.9)'')$$

subject to (10.8). As defined, $\underset{\sim}{\overline{\Phi}}$ has a pole $\overline{z}(c)$, so that

$$(10.11) \qquad (\overline{\Phi}^1(c, a), \ldots, \overline{\Phi}^n(c, a)) \equiv (c, 0, \ldots, 0) \qquad (a \in \beta_m).$$

Together (10.10) and (10.11) imply that the m-tuple $Y(c, a) = \underset{\sim}{0}$ for $a \in \beta_m$.

<u>The mate</u> $\underset{\sim}{Y}$ <u>of</u> $\underset{\sim}{\bar{\Phi}}$. For each m-tuple $\alpha \in \beta_m$, g^α is an extremal of J_f of the form (10.7)". We term the family $\underset{\sim}{Y}$ of these x-parameterized extremals g^α of J_f, the x-<u>parameterized mate</u> of $\underset{\sim}{\bar{\Phi}}$. It has just been seen that the extremals g^α of $\underset{\sim}{Y}$ meet the point $x = c$ on the x-axis in E_m.

To formulate Theorem 10.2 we set

$$(10.12) \qquad D_Y^c(x) = \frac{D(Y_1, \ldots, Y_m)}{D(\alpha_1, \ldots, \alpha_m)}(x, \alpha)\Bigg|^{\alpha = \underset{\sim}{0}} \qquad (0 \le x \le s_1)$$

and denote the corresponding Jacobian matrix by $\underset{\sim}{D}_Y^c(x)$. Theorem 10.2 relates the families $\underset{\sim}{Y}$ and $\underset{\sim}{\bar{\Phi}}$.

<u>Theorem</u> 10.2. <u>Let</u> $\underset{\sim}{\bar{\Phi}}$ <u>be a proper polar family of extremal arcs of</u> J_F <u>whose central extremal is the axial extremal</u> $\underset{\sim}{\bar{z}}$ <u>of Definition</u> 10.1 <u>and whose pole is the point</u> $\bar{z}(c)$ <u>of</u> $\underset{\sim}{\bar{z}}$. <u>Let</u> $\underset{\sim}{Y}$ <u>be the family of extremals</u> g^α <u>of</u> J_f <u>which are mates in</u> E_n <u>of extremals</u> Γ_α <u>of</u> $\underset{\sim}{\bar{\Phi}}$ <u>for which</u> $\|\alpha\|$ <u>is sufficiently small. Then the Jacobian</u> $D_Y^c(x)$ <u>of</u> (10.12) <u>and the Jacobian</u> $D_{\bar{\Phi}}^c(s)$ <u>of</u> (9.5) <u>are such that</u>

$$(10.13) \qquad D_{\bar{\Phi}}^c(s) \equiv D_Y^c(s) \qquad (0 \le s \le s_1)$$

<u>and the nullities of the corresponding functional matrices are equal for each</u> $s \in [0, s_1]$.

The mapping Y is defined by (10.10), subject to the condition (10.8). We shall make this definition more explicit in an equivalent form (10.15).

Let Z denote the ensemble of pairs (x, α) with $x \in K_\alpha$ where K_α is given by (10.9). For $(x, \alpha) \in Z$, condition (10.8) admits a unique solution

$s = S(x, \alpha)$ of class C^∞ such that $0 \leq s \leq s_1$. The central extremal arc of the family $\bar{\underset{\sim}{\Phi}}$ is the axial extremal $\bar{\underset{\sim}{z}}$ along which (10.3) holds. Hence

(10.14) $$\bar{\Phi}^1(s, \underset{\sim}{0}) \equiv s \qquad (0 \leq s \leq s_1) .$$

The condition (10.10) is equivalent to the identity

(10.15) $$[x, Y_1(x, \alpha), \ldots, Y_m(x, \alpha)] \equiv [\bar{\Phi}^1(S(x, \alpha), \alpha), \ldots, \bar{\Phi}^n(S(x, \alpha), \alpha)]$$

for $(x, \alpha) \in Z$.

In order to use the identity (10.15) to establish (10.13), the Jacobian of the mapping

(10.16) $$(x, \alpha) \to (S(x, \alpha), \alpha) : Z \to R^n ,$$

evaluated when $\alpha = \underset{\sim}{0}$ and $x \in [0, s_1]$, will be denoted by $j(x)$. The corresponding $n \times n$ functional matrix will be denoted by $\underset{\sim}{j}(x)$. On taking account of the identity $S(x, \underset{\sim}{0}) \equiv x$, one sees that $j(x) \equiv 1$. The matrix identity

(10.17) $$\left\| \begin{matrix} 1 , & \underset{\sim}{0} \\ \underset{\sim}{0} , & \underset{\sim}{D}^c_Y(x) \end{matrix} \right\| \equiv \underset{\sim}{D}^c_{\bar{\Phi}}(x)\, \underset{\sim}{j}(x) \qquad (0 \leq x \leq s_1)$$

is a consequence of (10.15). The two members of (10.17) are the Jacobian matrices of the two members of the identity (10.15) with respect to the variables $(x, \alpha_1, \ldots, \alpha_n)$, with evaluation when the m-tuple $\alpha = 0$.

Since $j(x) \equiv 1$, (10.13) follows from (10.17). Th concluding statement of Theorem 10.2 follows also from the matrix identity (10.17). The matrix

$\underset{\sim}{j}(x)$ is nonsingular and one makes use of the law of Sylvester as in the proof of Corollary 9.1.

Theorems on conjugate points. The meaning of the determinant $D_Y^c(x)$ and the corresponding matrix $\underset{\sim}{D}_Y^c(x)$ is made clear by the following corollary of Theorem 10.2 and Lemma 9.1.

Corollary 10.1. *In the sense of the nonparametric theory, the conjugate points of the point* $x = c$ *on the axial extremal* $\underset{\sim}{\bar{z}}$ *of* J_F *have multiplicities given by the nullities of the matrix* $\underset{\sim}{D}_Y^c(x)$ *at the points* $x \in [0, s_1]$ *other than* $x = c$, *at which* $D_Y^c(x) = 0$.

The mappings $x \to Y(x, a)$ of (10.7)'' for $a \in \beta_m$ are extremals of J_f, as has been seen. The columns of the Jacobian $D_Y^c(x)$ are n-tuples $Y_{a_\mu}(x, a)$, evaluated when $a = \underset{\sim}{0}$. According to Jacobi (see Mores [1], p. 25) these columns are solutions of the JDE[†] based on the axial mate $\underset{\sim}{g}$ of $\underset{\sim}{\bar{z}}$. Moreover, these solutions of the JDE all vanish when $x = c$. They are linearly independent. Otherwise $D_Y^c(x)$ would vanish identically, contrary to the identity $D_{\Phi}^c(s) \equiv D_Y^c(s)$ of (10.13), and the fact that $D_{\Phi}^c(x)$ does not vanish identically (Lemma 9.1).

In the nonparametric theory the conjugate point of the point $x = c$ on the axial extremal $\underset{\sim}{\bar{z}}$ of J_f are zeros of $D_Y^c(x)$, other than $x = c$. The multiplicity of such a conjugate point $x = c^*$ is (by definition in the nonparametric theory) the number of linearly independent solutions of the JDE which vanish at $x = c$ and at $x = c^*$. This number is obviously the nullity of $\underset{\sim}{D}_Y^c(c^*)$.

[†] JDE abbreviates "Jacobi differential equations."

Thus Corollary 10.1 is true.

The following corollary of Theorems 10.1 and 10.2 enables one to carry over properties of conjugate points known to be true in the nonparametric theory, to properties of conjugate points in the parametric theory.

Introduction to Corollary 10.2. Corresponding to an arbitrary presentation $(\hat{\psi}, \hat{V}) \in \mathcal{D}M_n$ and associated preintegrand G, let an extremal arc $\underset{\sim}{w}$ of J_G be given in the form (10.2)''. It follows from Theorem 22.1 that the following is true.

If V is a sufficiently small open neighborhood of $|\underset{\sim}{w}|$ in $\hat{V}$ and (ψ, V) the presentation in $\mathcal{D}M_n$ for which $\psi = \hat{\psi}|V$, then (ψ, V) is equivalent (as in Theorem 10.1) to a presentation (ϕ, U) under a transition diff $\Theta = \phi^{-1} \circ \psi$ such that $\Theta \cdot \underset{\sim}{w}$ is an axial extremal $\underset{\sim}{\bar{z}}$ in U, conjugatewise equivalent to $\underset{\sim}{w}$. As in Theorem 10.1, $\underset{\sim}{\bar{z}}$ is an extremal of the W-integral J_F associated with (ϕ, U).

The extremal arcs $\underset{\sim}{w}, \underset{\sim}{\bar{z}}$ and the x-parameterized extremal $\underset{\sim}{g}$ of[†] J_f which is the mate of $\underset{\sim}{\bar{z}}$ are conjugatewise equivalent in the following sense.

Corollary 10.2. On conjugatewise equivalence. A point $s = c^*$ on the extremal arc $\underset{\sim}{w}$ and on its axial equivalent $\underset{\sim}{\bar{z}} = \Theta \cdot \underset{\sim}{w}$, is a conjugate point of the point $s = c$ on $\underset{\sim}{w}$ or $\underset{\sim}{\bar{z}}$ and has a multiplicity μ, if and only if, in the sense of the nonparametric theory, the point $x = c^*$ is a conjugate point of the point $x = c$ and has the multiplicity μ on the x-parameterized mate $\underset{\sim}{g}$ of $\underset{\sim}{\bar{z}}$.

[†]Here f is the Euler mate of F defines in (8.3).

By virtue of Corollary 10.2 the following theorems are implied by theorems in the nonparametric theory.

Theorem 10.3. *On an extremal arc* $\underset{\sim}{z}$, *conjugate points of a point* $s = c$ *are isolated. The first conjugate point of a point* $s = c$ *on* $\underset{\sim}{z}$, *following or preceding* $s = c$ *on* $\underset{\sim}{z}$, *varies continuously strictly in the same sense as* c, *as long as such a first conjugate point exists on* $\underset{\sim}{z}$.

Notation for the Separation Theorem. The point $s = c$ on an extremal arc $\underset{\sim}{z}$ will be called an *improper conjugate* point of itself of multiplicity $n-1$. Let I be an arbitrary subinterval of the domain $[0, s_1]$. Let N_I^c denote the count† of proper or improper conjugate points in I of the point $s = c$. With this understood the following theorem is a consequence of the Extended Sturm Separation Theorem 20.1 of Morse [1] and of Corollary 10.2 of this book.

Separation Theorem 10.4. *Let* $s = c'$ *and* $s = c''$ *be two distinct points on an extremal arc* $\underset{\sim}{z}$ *on which* $0 \leq s \leq s_1$. *The multiplicity* ρ *(possibly 0) of* $s = c''$ *as a conjugate point of* $s = c'$ *equals the multiplicity of* $s = c'$ *as a conjugate point of* $s = c''$. *If* I *is an arbitrary subinterval of the domain* $[0, s_1]$ *of* $\underset{\sim}{z}$, *then*

$$|N_I^{c''} - N_I^{c'}| \leq m - \rho \qquad (m = n - 1) . \tag{10.18}$$

†Counting each conjugate point with its multiplicity.

Examples exist in which ρ is prescribed among integers on the range $0, 1, \ldots, n-1$. Examples exist in which $N_I^{c''} - N_I^{c'}$ is prescribed on the range $0, \pm 1, \pm 2, \ldots, \pm n-\rho$. The multiplicity of a conjugate point is at most $n-1$.

The following theorem is related to Theorem 10.4. It is a consequence of Corollary 20.1 of Morse [1] and of Corollary 10.2 of this book.

Theorem 10.5. *Under the hypotheses of Theorem 10.4 suppose that $c' < c''$. The count of conjugate points of c' on (c', c'') then equals the count of conjugate points of c'' on (c', c'').*

We shall state a theorem on the uniform absence of conjugate points. Theorem 10.6 is concerned with the W-integral J_F on a coordinate domain U_ϕ.

Theorem 10.6. *Let $\underset{\sim}{z}$ be an extremal arc of J_F on U_ϕ of R-length s_1 on which there are no points conjugate to the initial point $z(0)$ of $\underset{\sim}{z}$. There will then be no conjugate points on extremal arcs $\underset{\sim}{u}$ of J_F of R-length s_1 with initial elements $(u(0), \dot{u}(0))$ in $U \times \dot{R}^n$ such that*

$$\|u(0) - z(0)\| + \|\dot{u}(0) - \dot{z}(0)\| \tag{10.19}$$

*is sufficiently small.**

Corollary 10.2 on "conjugatewise equivalence," shows that Theorem 10.6 is a consequence of Theorems 5.1 of Morse [1], pp. 28-29. Theorems 5.1 of Morse [1] deal with a class of nonparametric integrals which include the "mates," in the sense of §8, of the W-integrals J_F of Theorem 10.6.

---

* We take the value of (10.19) as a measure of the "phase-wise" nearness of $\underset{\sim}{u}$ and $\underset{\sim}{z}$.

§11. The measure of degenerate extremal joins. Let (P_0, P) be a pair of distinct points on M_n joined by an extremal γ of J with an initial point P_0 and terminal point P. If γ is simple, conjugate points of P_0 on γ are well-defined in §10. If γ is not simple, conjugate points of P_0 on γ will be defined in §23. If P_0 is conjugate to P on γ, γ will be called a degenerate extremal join of P_0 to P. Fixing P_0, we shall measure the set of degenerate extremal joins with pole P_0 by measuring the set of points P on M_n which are conjugate to P_0 on some extremal γ joining P_0 to P.

Sets of measure 0 on M_n. Let X be a set of points on M_n included in the range $\phi(U)$ of a presentation $(\phi, U) \in \mathcal{D}M_n$. The set X will be said to have a measure $m(X) = 0$, on M_n if the set $\phi^{-1}(X)$ has the Lebesgue measure 0 on U. If X is included in the range of a second presentation $(\psi, V) \in \mathcal{D}M_n$, the Lebesgue measure of $\psi^{-1}(X)$ on V will be zero if and only if the Lebesgue measure of $\phi^{-1}(X)$ is 0 on U. More generally, an arbitrary subset Z of points on M_n can be included in the union of a countable set of ranges of presentations in $\mathcal{D}M_n$. The set Z will be said to have a measure 0 on M_n if its intersection with the range of each presentation in $\mathcal{D}M_n$ has the measure 0.

The proof of the following theorem will be completed in §24, after defining conjugate points on self-intersecting extremals of J.

Measure[†] Theorem 11.1. Corresponding to a point A_1, prescribed

---

[†]A theorem similar to Theorem 11.1 was proved as Theorem 12 on p. 233 of my Colloquim Lectures of 1932. The earlier theorem was based on an equivalent definition of conjugate points by means of the Jacobi differential equations in tensor form. In this book conjugate points are defined by the zeros of Jacobians of proper polar families of extremals.

<u>and fixed in</u> M_n, <u>the set of points</u> P <u>on</u> M_n <u>such that</u> A_1 <u>is joined to</u> P <u>by a degenerate extremal of</u> J, <u>has the measure</u> 0 <u>on</u> M_n.

We shall begin the proof of Theorem 11.1 by proving Lemma 11.1 and its corollary.

<u>Introduction to Lemma</u> 11.1. Let (ϕ, U) be a presentation in $\mathcal{D}M_n$ and J_F the associated local W-integral. Let $\underset{\sim}{\Phi}$ be a proper polar family of extremal arcs of J_F in U, with pole u_0 when $s = 0$. Conjugate points of u_0 are defined, as in §10, on the <u>central</u> extremal arc $\underset{\sim}{z}$ of $\underset{\sim}{\Phi}$. Lemma 11.1 shows how to determine the conjugate points of u_0 on <u>any</u> extremal arc of $\underset{\sim}{\Phi}$ sufficiently near $\underset{\sim}{z}$.

<u>Lemma</u> 11.1. <u>If</u> e <u>is a sufficiently small positive constant, the conjugate points of</u> $s = 0$ <u>on extremals</u> $\Gamma_a = \Phi(\cdot, a)$ <u>of the family</u> $\underset{\sim}{\Phi}$ <u>for which</u> $\|a\| < e$, <u>are the points</u> $u = \Phi(s, a) \in U$ <u>on</u> Γ_a <u>at which</u> $0 < s \leq s_1$ <u>and the Jacobian</u>

$$(11.1)^{\dagger} \qquad D_{\Phi}(s, a) = \frac{D(\Phi^1, \Phi^2, \ldots, \Phi^n)}{D(s, a_1, \ldots, a_m)}(s, a) = 0 .$$

The zeros of $D_{\Phi}(s, \underset{\sim}{0})$ when $0 < s \leq s_1$ are, by Definition 10.0(i), the conjugate points of $s = 0$ on $\Gamma_{\underset{\sim}{0}}$. We must show that the conjugate points s of $s = 0$ on Γ_a are still given by the condition (11.1), when $a \neq \underset{\sim}{0}$, provided $\|a\|$ is sufficiently small.

[†] In the notation of §9 and §10

$$D_{\Phi}(s, \underset{\sim}{0}) = D^0_{\Phi}(s) .$$

To that end suppose that e is so small that the extremals Γ_α are well-defined for $\|\alpha\| < 2e$ and that the functional matrix (9.4) remains of rank $n-1$ when $s = c = 0$ and $\|\alpha\| < 2e$. To determine conjugate points on Γ_α for $\alpha \neq \underset{\sim}{0}$ in accord with Definition 10.0(i), we must fix α and represent Γ_α as the "central" extremal of a proper polar family $\underset{\sim}{\Psi}$.

<u>Definition of</u> $\underset{\sim}{\Psi}$. Let β be an m-tuple in R^m. For α fixed, with $\|\alpha\| < e$, set

$$\Psi(s, \beta) = \Phi(s, \beta+\alpha) \qquad (0 \leq s \leq s_1, \ \|\beta\| < e) . \tag{11.2}$$

For $\|\beta\| < e$, the extremals,

$$s \to \Psi(s, \beta) : [0, s_1] \to U \tag{11.3}$$

form a proper polar sub-family $\underset{\sim}{\Psi}$ of extremals of $\underset{\sim}{\Phi}$, with Γ_α as central extremal $\Psi(\cdot, \underset{\sim}{0})$ and with pole the point $s = 0$ on these extremals. Moreover for the fixed α and $0 \leq s \leq s_1$,

$$\frac{D(\overset{1}{\Psi}, \overset{2}{\Psi}, \ldots, \overset{n}{\Psi})}{D(s, \beta_1, \ldots, \beta_m)} (s, \underset{\sim}{0}) = \frac{D(\overset{1}{\Phi}, \overset{2}{\Phi}, \ldots, \overset{n}{\Phi})}{D(s, \alpha_1, \ldots, \alpha_m)} (s, \alpha) . \tag{11.4}$$

By Definition 10.1 the zeros $s > 0$ of the Jacobian $D^0_\Psi(s)$ on the left of (11.4) give the conjugate points of $s = 0$ on $\Psi(\cdot, \underset{\sim}{0})$. Since $\Psi(\cdot, \underset{\sim}{0}) = \Gamma_\alpha$, Lemma 11.1 follows.

A basic corollary of Lemma 11.1 will now be stated.

<u>Corollary</u> 11.1. <u>For</u> e <u>conditioned as in Lemma</u> 11.1, <u>let</u> X_e <u>be the set of conjugate points in</u> U <u>of</u> $s = 0$, <u>on extremals</u> $\Gamma_\alpha = \Phi(\cdot, \alpha)$ <u>of</u> J_F

for which $\|a\| < e$. *The set* X_e *then has the measure* 0 *on* U.

By virtue of Lemma 11.1,

$$X_e = \{ u = \Phi(s, a) \| 0 < s \leq s_1, \ \|a\| < e; D_\Phi(s, a) = 0\} \ .$$

According[†] to "Sard's Theorem," X_e is a set of Lebesgue measure 0 on U. Thus Corollary 11.1 is true.

[†] Corollary 11.1 is similarly implied by Lemma 11.1 and by Lemma 6.3 on page 41 of Morse-Cairns. The latter lemma was proved by Morse in 1926, but not then published. Instead its concepts were incorporated in the proof of Lemma 12.1 of Morse [1], page 231, on the nullity of the measure of the conjugate points of a point P on extremals issuing from P.

Part III

Minimizing arcs

Chapter 5

Necessary conditions

§12. Necessity of the Euler condition. Let F be a W-preintegrand associated with a presentation (ϕ, U) in $\mathcal{D}M_n$. Concerning F a classical theorem will be proved.

Theorem 12.1. Let

(12.1)
$$ \underset{\sim}{z} : t \to z(t) : [t_0, t_1] \to U $$

be a regular arc on U. If $J_F(\underset{\sim}{z}) \leq J_F(\underset{\sim}{u})$ for each piecewise regular arc $\underset{\sim}{u}$ joining the end points of $\underset{\sim}{z}$ in some open neighborhood N of $\underset{\sim}{z}$ in U, then for $t_0 \leq t \leq t_1$

(12.2)
$$ \frac{d}{dt} F_{r_i}(z(t), \dot{z}(t)) = F_{u^i}(z(t), \dot{z}(t)) \qquad (i = 1, \ldots, n) . $$

Proof. For $i = 1, \ldots, n$ let $t \to \eta^i(t) : [t_0, t_1] \to R$ be a mapping of class D^1 such that $\eta^i(t_0) = \eta^i(t_1) = 0$. If e is a sufficiently small positive constant, the mapping

(12.3)
$$ t \to z(t) + e\eta(t) : [t_0, t_1] \to R^n $$

is a piecewise regular curve of class D^1, joining the end points of $\underset{\sim}{z}$ in N. By hypothesis

(12.4)
$$ \int_{t_0}^{t_1} F(z(t), \dot{z}(t)) \, dt \leq \int_{t_0}^{t_1} F(z(t) + e\eta(t), \dot{z}(t) + e\dot{\eta}(t)) \, dt . $$

The derivative as to e of the integral on the right of (12.4) accordingly vanishes when $e = 0$. Thus

(12.5)
$$\int_{t_0}^{t_1} (F^t_{r^i}\dot{\eta}^i(t) + F^t_{u^i}\eta^i(t))\, dt = 0$$

where the superscript t indicates evaluation for $(u, r) = (z(t), \dot{z}(t))$. The condition (12.5) implies that

(12.6)
$$\int_{t_0}^{t_1} \dot{\eta}^i(t)\left(F^t_{r^i} - \int_{t_0}^{t} F^a_{u^i}\, da\right) dt = 0$$

for each admissible η. By virtue of the DuBois-Raymond Lemma

(12.7)
$$F^t_{r^i} = \int_{t_0}^{t} F^a_{u^i}\, da + c_i \qquad (i = 1, \ldots, n)$$

for suitable constants c_i. The relations (12.2) follow on differentiating the members of (12.7) as to t.

§13.. The Weierstrass necessary condition. Let a presentation (ϕ, U) be given with an associated W-preintegrand F. Let $\mathcal{E}_F$ be the Weierstrass function associated with F in (5.16). The principal theorem of this section follows.

Theorem 13.1. *If the regular arc $\underset{\sim}{z}$ of (12.1) affords a relative minimum to J_F in the sense of Theorem 12.1, then*

$$\mathcal{E}_F(z(t), \dot{z}(t), r) \geq 0 \qquad (t_0 \leq t \leq t_1) \tag{13.1}$$

for arbitrary nonnull n-tuples r.

We can suppose that $\underset{\sim}{z}$ is reparameterized so that $t_0 = 0$ and $t_1 > 1$. Such a reparameterization will not affect the value of J_F, nor will it affect the validity of (13.1), because of the homogeneity of $\mathcal{E}_F$ as formulated in (5.20).

Understanding that $t_0 = 0$ and $t_1 > 1$ we shall show that for arbitrary nonnull n-tuples r

$$\mathcal{E}_F(z(t), \dot{z}(t), r) \geq 0 \qquad (0 < t < t_1) . \tag{13.2}$$

It will follow from continuity considerations that the condition on $\mathcal{E}_F$ in (13.2) holds when $t = 0$ or t_1. It will be sufficient to prove that

$$\mathcal{E}_F(z(1), \dot{z}(1), r) \geq 0 \qquad (\text{for } r \neq \underset{\sim}{0}) , \tag{13.3}$$

because a suitable reparameterization of $\underset{\sim}{z}$, holding $t = 0$ and $t = t_1$ fast, will carry a point t, prescribed in $(0, t_1)$, into $t = 1$.

To prove (13.3) we shall compare the value of $J_F(\underset{\sim}{z})$ with the value of J_F along a continuous curve k_a, joining the end points of $\underset{\sim}{z}$, defined for each value of a parameter $a \leq 1$ and converging uniformly[†] to $\underset{\sim}{z}$ as a converges to 1. The curve k_a will be defined by a sequence of three regular arcs with independent parameterizations.

The third arc of k_a shall be the subarc of $\underset{\sim}{z}$ on which $1 \leq t \leq t_1$. This arc is independent of a. Along this arc J_F is a constant K independent of a.

The second arc of k_a shall be a short regular arc $a \to u(a)$ in U, defined for values of $a \leq 1$ near $a = 1$. It shall terminate with the point $u(1) = z(1)$ and be such that $\dot{u}(1) = r$, where r is a prescribed nonnull n-tuple independent of a. Taken along this arc from an initial point $u(a)$ to the terminal point $u(1) = z(1)$, J_F is a function of a whose derivative when $a = 1$ is $-F(z(1), r)$.

The first arc of k_a must join the initial point $z(0)$ of $\underset{\sim}{z}$ to the initial point $u(a)$ of the second arc of k_a. The first arc of k_a is defined by a mapping

$$t \to q_a(t) : [0,1] \to U \tag{13.4}$$

where, for $a \leq 1$ and a near 1,

$$q_a(t) = z(ta) + (u(a) - z(a))t \qquad (0 \leq t \leq 1) . \tag{13.5}$$

[†]When k_a is reparameterized by R-length.

Note that $q_1(t) \equiv z(t)$, that $q_a(0) = z(0)$ and that $q_a(1) = u(a)$. It follows that k_a joins the end points of $\underset{\sim}{z}$. Note that k_1 reduces to $\underset{\sim}{z}$.

The partial derivative of $q_a^i(t)$ as to a, if evaluated when $a = 1$ and $t = 1$, reduces to $\dot{u}^i(1) = r^i$. Taken along the first arc of k_a, J_F is a function of a whose derivative as to a, evaluated when $a = 1$, is $\dot{u}^i(1)F_{r^i}(z(1), \dot{z}(1))$, *as one sees by an appropriate integration by parts.*

By hypothesis $J_F(\underset{\sim}{z}) \leq J_F(k_a)$. Hence

$$\frac{d}{da}J_F(k_a)\Big|^{a=1} \leq 0 . \tag{13.6}$$

Our calculations in the preceding paragraphs, taken with (13.6), show that

$$F(z(1), r) - \dot{u}^i(1)F_{r^i}(z(1), \dot{z}(1)) \geq 0 . \tag{13.7}$$

Since the n-tuple $\dot{u}(1)$ equals the prescribed n-tuple r, (13.3) follows from (13.7) and (5.16).

Thus Theorem 13.1 is true.

§14. Weak minima and positive regularity. Let (ϕ, U) be a presentation in $\mathcal{D}M_n$ and F the associated W-preintegrand. Let

(14.1)
$$\underset{\sim}{z} : t \to z(t) : [t_0, t_1] \to U$$

be a regular arc in U. To characterize conditions under which $\underset{\sim}{z}$ affords a weak minimum to J_F when the boundary conditions fix the endpoints, notational innovations are needed.

The phase space $U \times \dot{R}^n$. The subspace $U \times \dot{R}^n$ of R^{2n} of 2n-tuples

$$(u^1, \ldots, u^n : r^1, \ldots, r^n)$$

in which $u \in U$ and $\| r \| \neq 0$, will be called the phase space associated with U. By the phase image $Ph \underset{\sim}{z}$ of the regular arc $\underset{\sim}{z}$: (14.1) is meant the curve

(14.2)
$$t \to (z(t), \dot{z}(t)) : [t_0, t_1] \to U \times \dot{R}^n .$$

The open e-neighborhood in $U \times \dot{R}^n$ of the carrier of $Ph \underset{\sim}{z}$ will be termed the e-tube of $Ph \underset{\sim}{z}$ in $U \times \dot{R}^n$. When $\underset{\sim}{u}$ is piecewise regular $Ph \underset{\sim}{u}$ will be taken as the sequence of phase images in $U \times \dot{R}^n$ of the regular curves which define $\underset{\sim}{u}$. If the curve $\underset{\sim}{u}$ has μ corners, $Ph \underset{\sim}{u}$ will have μ discontinuities.

Definition 14.1. A weak relative minimum of J_F. A regular arc $\underset{\sim}{z}$ of form (14.1) in the domain U of J_F will be said to afford a weak relative minimum to J_F in the fixed endpoint problem, if for a sufficiently small positive constant e

$$J_F(\underset{\sim}{z}) \leq J_F(\underset{\sim}{u}) \tag{14.3}$$

for each piecewise regular arc $\underset{\sim}{u}$ joining the endpoints of $\underset{\sim}{z}$ in U and such that Ph $\underset{\sim}{u}$ is included in the e-tube of Ph $\underset{\sim}{z}$.

An examination of the proof of Theorem 12.1 discloses the truth of the following.

__Theorem__ 14.1. _If a regular arc_ $\underset{\sim}{z}$: (14.1) _in_ U _affords a weak relative minimum to_ J_F _in the sense of Definition_ 14.1, _then_ $\underset{\sim}{z}$ _satisfies the Euler equations of_ J_F.

The proof of Theorem 13.1 shows that the following is true.

__Theorem__ 14.2. _If a regular arc_ $\underset{\sim}{z}$: (14.1) _in_ U _affords a weak relative minimum to_ J_F, _then if_ $\varepsilon > 0$ _is sufficiently small_

$$\mathcal{E}_F(z(t), \dot{z}(t), r) \geq 0 \tag{14.4}$$

for each $t \in [t_0, t_1]$ _and_ n-_tuple_ r _such that_ $\|r - \dot{z}(t)\| < \varepsilon$.

We come now to a special necessary condition, a condition that will be called the condition of "positive regularity" of F along the curve $\underset{\sim}{z}$.

__Theorem__ 14.3. _If a regular arc_ $\underset{\sim}{z}$: (14.1) _in_ U _affords a weak relative minimum to_ J_F, _it is necessary that_

$$F_{r^i r^j}(z(t), \dot{z}(t))\lambda^i \lambda^j > 0 \tag{14.5}$$

for t _prescribed in_ $[t_0, t_1]$ _and for each_ n-_tuple_ λ _not a scalar multiple of_ $\dot{z}(t)$.

Let r be an arbitrary nonnull n-tuple. Fixing t, consider the mapping

$$e \to \theta(e) = \mathcal{E}_F(z(t), \dot{z}(t), \dot{z}(t) + er) \tag{14.6}$$

for values of $|e|$ so small that $\dot{z}(t) + er \neq \underset{\sim}{0}$. Observe that $\theta(0) = 0$. According to Theorem 14.2, $\theta(0)$ gives a relative minimum to θ. Hence $\theta'(0) = 0$ and $\theta''(0) \geq 0$. A computation of $\theta''(0)$ shows that

$$F_{r^i r^j}(z(t), \dot{z}(t)) r^i r^j \geq 0 \qquad (t_0 \leq t \leq t_1) \tag{14.7}$$

For the given fixed t, the equality will hold in (14.7) for a nonnull n-tuple r, if and only if

$$F_{r^i r^j}(z(t), \dot{z}(t)) r^i = 0 \qquad (j = 1, \ldots, n) \tag{14.8}$$

It follows from (4.4) that (14.8) holds if r is a scalar multiple of $\dot{z}(t)$. By virtue of Definition 6.1 of the "nonsingularity" of F, the rank of $\| F_{r^i r^j}(z(t), \dot{z}(t)) \|$ is n-1. It follows that (14.8) holds for $r \neq \underset{\sim}{0}$, if and only if r is a scalar multiple of $\dot{z}(t)$. Equivalently

$$F_{r^i r^j}(z(t), \dot{z}(t)) \lambda^i \lambda^j = 0 \tag{14.9}$$

for a t prescribed in $[t_1, t_2]$, if and only if the n-tuple λ is a scalar multiple of the n-tuple $\dot{z}(t)$.

Theorem 14.3 follows.

The invariance of weak minima. Suppose that the presentation (ϕ, U) is equivalent to the presentation (ψ, V) in that $\phi(U) = \psi(V)$, so that the transition diff $\theta = \psi^{-1} \circ \phi$ maps U onto V. W-preintegrands F and G are associated, respectively, with the presentations (ϕ, U) and (ψ, V). Corresponding to the regular arc $\underset{\sim}{z}$ of (14.1) the image $\underset{\sim}{w}$ of $\underset{\sim}{z}$ under θ is a regular arc

$$\underset{\sim}{w} : t \to w(t) : [t_0, t_1] \to V \tag{14.10}$$

The compatibility of the W-preintegrands F and G implies that $\underset{\sim}{z}$ affords a weak minimum to J_F if and only if $\underset{\sim}{w} = \theta \cdot \underset{\sim}{z}$ affords a weak minimum to J_G.

From Theorem 14.3 we infer the following. If $\underset{\sim}{w}$ affords a weak minimum to J_G then

$$G_{\dot{\sigma}^i \dot{\sigma}^j}(w(t), \dot{w}(t))\mu^i \mu^j > 0 \tag{14.11}$$

for each $t \in [t_0, t_1]$ and nonnull n-tuple μ not a scalar multiple of $\dot{w}(t)$.

The necessary condition of Theorem 14.3 is called the condition of positive regularity of F along the extremal $\underset{\sim}{z}$ and leads to the following definition.

Definition 14.2. Positive regularity of F. A W-preintegrand F is said to be Pos-R* at a 2n-tuple (u, r) in its domain $U \times \dot{R}^n$, if

$$F_{\dot{r}^i \dot{r}^j}(u, r)\lambda^i \lambda^j > 0, \tag{14.12}$$

* Pos-R abbreviates positive regular or positive regularity, whichever phrase is relevant in the given context.

for each nonnull n-tuple λ not a scalar multiple of r. The preintegrand F is said to be Pos-R if it is Pos-R at each 2n-tuple (u, r) in its domain.

The condition of Pos-R may be regarded as evolving from the classical condition of Legendre when $m = 1$.

Definition 14.3. *Positive regularity of* J. The W-integral J on M_n will be said to be Pos-R if each W-preintegrand F is Pos-R.

According to Theorem 15.1, if at least one W-preintegrand F is Pos-R, J is Pos-R. An even simpler condition for positive regularity of J is implied by Theorem 18.2.

Exercise 14.1. Let F be a W-preintegrand of a presentation $(\phi, u) \in \mathcal{L}M_n$. Let f be the "Euler mate" of F with values $f(x, y, p)$ defined as in (8.3). Prove the following lemma.

Lemma 14.1. *Let* $\underset{\sim}{w}$ *be an extremal*

$$s \to w(s) : [0, s_1] \to U \tag{14.13}$$

of J_F *on which* $u^1 \equiv s$ *and whose carrier is identical with the segment* $0 \leq u^1 \leq s_1$ *of the* u^1*-axis. If* F *is* Pos-R *along* $\underset{\sim}{w}$, *the quadratic form whose coefficients are given by the* (n-1)*-square matrix*

$$\| f_{p_\mu p_\nu}(x, \underset{\sim}{0}, \underset{\sim}{0}) \| \qquad (\mu, \nu = 1, 2, \ldots, m) \tag{14.14}$$

is positive definite for $0 \leq x \leq s_1$.

Suggestion. Review the proof of Lemma 8.1 and show that each element in the first row and column of $\| F_{r^i r^j}(w, \dot{w}) \|$ vanishes.

§15. The positive regularity of W-preintegrands. Theorem 15.1 is believed to be new.

Theorem 15.1. On positive regularity. Under the hypothesis of nonsingularity of each W-preintegrand, the positive regularity of one such W-preintegrand at one 2n-tuple (u_0, r_0) in its domain implies the positive regularity of each W-preintegrand at each 2n-tuple in its domain.

The proof of Theorem 15.1 depends on three lemmas.

Lemma 15.1. Let F be a W-preintegrand associated with a presentation (ϕ, U) such that U is connected. If F is Pos-R at a 2n-tuple (u_0, r_0) in the domain of F, it is Pos-R at each 2n-tuple (u_1, r_1) in the domain of F.

The n-square matrix $K_0 = \| F_{r^i r^j}(u_0, r_0) \|$ has just one characteristic root which is zero; for the matrix K_0 has the rank n-1 by virtue of the hypothesis of nonsingularity of each W-preintegrand. The remaining n-1 characteristic roots of K_0 are positive, since F is Pos-R at (u_0, r_0) by hypothesis. Since U is connected by hypothesis, a 2n-tuple (u, r) can be varied continuously in the domain of F from (u_0, r_0) to (u_1, r_1). During this variation the rank of $\| F_{r^i r^j}(u, r) \|$ will remain constantly n-1 so that one and only one characteristic root of $\| F_{r^i r^j}(u_1, r_1) \|$ will vanish while the other characteristic roots will be positive.

Lemma 15.1 follows.

Lemma 15.2 (i). Let F and G be W-preintegrands associated with equivalent presentations, respectively, (ϕ, U) and (ψ, V). Then F is

Pos-R over the domain of F, if and only if G is Pos-R over the domain of G.

(ii). *Let F and G be W-preintegrands associated, respectively, with presentations (ϕ, U) and (ψ, V) such that $V \subset U$ and $\psi = \psi | V$. Then G is Pos-R at a 2n-tuple (v, σ) in the domain of G, if and only if F is Pos-R at (v, σ).*

Proof of (i). Let θ be the transition diff $\psi^{-1} \circ \phi$ of U onto V. If a 2n-tuple (u, r) in the domain of F and a 2n-tuple (v, σ) in the domain of G are related in that $v = \theta(u)$ and σ is the contravariant image at v of r at u, then Lemma 5.2 affirms that

(15.1) $$F_{r^i r^j}(u, r)\lambda^i \lambda^j = G_{\sigma^i \sigma^j}(v, \sigma)\mu^i \mu^j$$

for each n-tuple λ at u and contravariant image μ at v of λ at u. It follows that F is Pos-R at (u, r) if and only if G is Pos-R at the related 2n-tuple (v, σ). As (u, r) ranges over the domain of F the related 2n-tuples (v, σ) will range over the domain of G. Thus (i) is true.

Proof of (ii). By the compatibility condition I on F and G of §3, $F(v, r) = G(v, \sigma)$ for each pair (v, σ) in the domain of G. Hence (ii) is true.

This completes the proof of Lemma 15.2.

Lemma 15.3. *Let F and G be W-preintegrands associated, respectively, with presentations (ϕ, U) and (ψ, V), such that U and V are connected sets* *and* $\phi(U)$ *meets* $\psi(V)$. *Then* F *is* Pos-R *on its domain if and only if* G *is* Pos-R *on its domain.*

Since $\phi(U)$ meets $\psi(V)$, there exist equivalent presentations $(\hat{\phi}, \hat{U})$ and $(\hat{\psi}, \hat{V})$ which are restrictions, respectively, of (ϕ, U) and (ψ, V). Let $\hat{F}$ and $\hat{G}$ be the W-preintegrands associated, respectively, with $(\hat{\phi}, \hat{U})$ and $(\hat{\psi}, \hat{V})$. By virtue of Lemma 15.2(ii) F is Pos-R over its domain if and only if $\hat{F}$ is Pos-R over its domain. By Lemma 15.1, $\hat{F}$ is Pos-R over its domain if and only if $\hat{G}$ is Pos-R over its domain. By Lemma 15.2(ii), $\hat{G}$ is Pos-R over its domain if and only if G is Pos-R over its domain.

Lemma 15.3 follows.

Completion of proof of Theorem 15.1. Let F and G be W-preintegrands associated respectively with presentations (ϕ, U) and (ψ, V). We seek to prove the following:

(a) *If* F *is* Pos-R *at a* 2n-*tuple* (u, r) *in its domain, then* G *is* Pos-R *at any* 2n-*tuple* (v, σ) *in its domain.*

By virtue of Lemma 15.2(ii) it will suffice to prove (a) under the assumption that U and V are each connected sets. Since M_n is connected there exists a sequence,

$$(15.2) \qquad (\phi_1, U_1), \ldots, (\phi_s, U_s) \qquad (s > 1)$$

of presentations in $\mathcal{D}M_n$ with the following three properties.

(a_1) $u \in U_1 \subset U$ and $v \in U_s \subset V$.

(a_2) The range on M_n of each of the presentations (15.2), except the first, meets the range on M_n of its predecessor.

(a_3) Each coordinate domain U_i is connected.

Let $F_1, \ldots, F_s$ be W-preintegrands associated with the respective presentations (15.2). By Lemma 15.3, F_1 is Pos-R on U_1 with F, F_2 is Pos-R on U_2 with F_1, F_3 with $F_2, \ldots, F_s$ with F_{s-1} and finally G with F_s.

Theorem 15.1 follows.

§16. The necessity of the Jacobi condition. Let an extremal arc $\underset{\sim}{w}$ be given as in (10.3) in the coordinate domain of (ψ, V). Theorem 16.1 formulates the Jacobi necessary condition.

Theorem 16.1. *If an extremal arc*

$$\underset{\sim}{w} : s \to w(s) : [0, s_1] \to V \tag{16.1}$$

affords a weak minimum to the W-integral J_G associated with a presentation (ψ, V) in the fixed endpoint problem, it is necessary that the initial point of $\underset{\sim}{w}$ be conjugate to no point on $\underset{\sim}{w}$ preceding the final endpoint of $\underset{\sim}{w}$.

Theorem 16.1 will be established as a simple consequence of the corresponding theorem in the nonparametric theory.

According to Corollary 10.2, a sufficiently small open neighborhood V_0 of $\underset{\sim}{w}$ in V can be mapped by a transition diff Θ onto a neighborhood U in E_n of an axial extremal $\underset{\sim}{\bar{z}}$ of a W-integral J_F. The x-parameterized mate $\underset{\sim}{g}$ of $\underset{\sim}{\bar{z}}$ is a solution of the Euler equations J_f, where f is the Euler mate of F, defined in (8.3). According to Corollary 10.2, the above extremals, $\underset{\sim}{w}$ of J_G, $\underset{\sim}{\bar{z}}$ of J_F and $\underset{\sim}{g}$ of J_f are "conjugatewise equivalent."

By hypothesis, $\underset{\sim}{w}$ affords a weak minimum to J_G. It follows that $\underset{\sim}{\bar{z}}$ affords a weak minimum to J_F. We infer from the equality (8.10) of the Euler and Weierstrass integrals, that $\underset{\sim}{g}$ affords a weak minimum* to J_f as well. Since the determinant

$$|f_{p_\mu p_\nu}(x, y, p)| \neq 0 \tag{16.2}$$

(cf. (8.13))

---

* The minimum afforded to J_f by $\underset{\sim}{g}$ is weak in the sense of the nonparametric theory. See p. 18 of Morse [1].

the "Jacobi differential equations" based on the extremal g of J_F are nonsingular. In the nonsingular case a classical theorem of the nonparametric theory implies that the initial point of the minimizing extremal g is conjugate to no point on g preceding the terminal point of g. Cf. Morse [1], Theorem 4.1.

Since the extremal w of J_G and the extremal g of J_f are "conjugatewise equivalent" in the sense of Corollary 10.2, Theorem 16.1 follows.

Chapter 6

Sufficient Conditions

§17. A Hilbert integral H_F. In Chapter 5 there is given a presentation (ϕ, U) and an associated W-integral J_F. An extremal arc

(17.0) $$\underset{\sim}{z} : s \to z(s) : [a, b] \to U$$

of J_F is given. This arc is R-parameterized in accord with Definition 7.1. The object of Chapter 5 is to establish sufficient conditions that $\underset{\sim}{z}$ afford a minimum to J_F relative to piecewise regular arcs $\underset{\sim}{u}$ which join the endpoints of $\underset{\sim}{z}$ in some neighborhood in U of the carrier $|\underset{\sim}{z}|$ of $\underset{\sim}{z}$.

A Hilbert integral H_F is a line integral on U associated with J_F. It is used in studying the conditions under which $\underset{\sim}{z}$ is a minimizing extremal. The definition of H_F presupposes the existence of a family Γ of extremals of J_F in U.

The representation of Γ. To each extremal of Γ an m-tuple β in R^m will be assigned as a parameter and the extremal denoted by Γ^β. The m-tuple β shall be a point on an open simply-connected subset B of R^m. To each point on Γ^β a parameter s will be assigned. The value s_0 of s at the initial points p_0 of extremals Γ^β will be arbitrarily prescribed. The value s at each other point p of an extremal Γ^β shall be such that $s - s_0$ is the R-length of Γ^β, measured from the initial point p_0 of Γ^β to the point p. The value of s at the terminal point of Γ^β will be denoted by s_1. The R-length of Γ^β is thus $s_1 - s_0$. We understand that s_0 and s_1 are independent of $\beta \in B$.

Set $\Pi = [s_0, s_1] \times B$. The family Γ is supposed given by a C^∞-mapping

$$(s, \beta) \to \Gamma(s, \beta) : \Pi \to U \tag{17.1}$$

whose partial mappings for fixed $\beta \in B$ define the respective extremals Γ^β. The domain Π of the mapping Γ is in R^n, but not open in R^n. We shall suppose that for some sufficiently small open interval I which includes $[s_0, s_1]$, the extremals of the "family" Γ admit extensions for $s \in I$ and that the "mapping" Γ admits a C^∞-extension over $I \times B$.

Γ as a field. Let $\hat{\Gamma}$ be the extension of Γ over $I \times B$. The extremals of Γ will be said to define a field in U if for some choice of the interval I the extension $\hat{\Gamma}$ of Γ is a diff* of $I \times B$ onto an open subset of U. Γ maps Π onto $\Gamma(\Pi)$. In U an n-tuple will be denoted by u.

The inverse of Γ is a C^∞-mapping of $\Gamma(\Pi)$ onto the subset Π of R^n. Γ^{-1} will be written in the form

$$u \to \big(s(u), \beta(u)\big) : \Gamma(\Pi) \to \Pi \ . \tag{17.2}$$

The n-tuple

$$\rho(u) = \Gamma_s\big(s(u), \beta(u)\big) \qquad (u \in \Gamma(\Pi)) \tag{17.3}$$

defines a vector tangent to the extremal $\Gamma^{\beta(u)}$ at the point u. It will be called the direction of the field Γ at u. The vector $\rho(u)$ is "R-unitary" in that $\mathcal{f}(u, \rho(u)) = 1$. See (2.6) for a definition of $\mathcal{f}$.

---

* We understand that a diff has a nonnull Jacobian at each of the points at which it is defined and that it is globally 1-1.

The Hilbert integral defined. If χ is any piecewise regular curve in the subset $\Gamma(\Pi)$ of U, the Hilbert integral $H_F(\chi)$ is defined by the line integral

(17.4) $$H_F(\chi) = \int_{\chi} F_{r^i}(u, \rho(u))\, du^i$$

Since Γ is a diff of Π onto $\Gamma(\Pi)$, the Hilbert integral can be represented by a line integral in the space of n-tuples $(s,\beta) \in \Pi$. To that end let λ be the piecewise regular curve on Π which is the image of χ under the diff Γ^{-1} of $\Gamma(\Pi)$ onto Π. The integral (17.4) can be written as a line integral

(17.5) $$H^*(\lambda) = \int_{\lambda} M(s,\beta)\, ds + N_{\mu}(s,\beta)\, d\beta_{\mu} \qquad (\mu = 1, \ldots, m)$$

A simple computation shows that for $(s,\beta) \in \Pi$

(17.6)' $$M(s,\beta) \equiv F\big(\Gamma(s,\beta), \Gamma_s(s,\beta)\big)$$

(17.6)'' $$N_{\mu}(s,\beta) \equiv F_{r^i}\big(\Gamma(s,\beta), \Gamma_s(s,\beta)\big)\Gamma^i_{\beta_{\mu}}(s,\beta) \qquad (\mu = 1, \ldots, m)$$

Verification of (17.6)' and (17.6)''. One starts with (17.4) and the identity $u \equiv \Gamma\big(s(u), \beta(u)\big)$, valid for $u \in \Gamma(\Pi)$. Subject to the diff Γ, one finds that

$$du^i = \Gamma^i_s(s,\beta)\, ds + \Gamma^i_{\beta_{\mu}}(s,\beta)\, d\beta_{\mu}$$

and that by virtue of (4.2) and (17.3)

$$F_{r^i}(u,\rho(u))\rho^i(u) = F\big(\Gamma(s,\beta),\Gamma_s(s,\beta)\big),$$

where $\rho(u)$ is given by (17.3). The relations (17.6) follow.

We shall verify the following theorem.

Theorem 17.1. *A necessary and sufficient condition that the Hilbert integral H_F be independent of admissible* paths joining two points in $\Gamma(\Pi)$ is that for μ, ν on the range $1, \ldots, m = n-1$*

(17.7) $$\frac{\partial}{\partial\beta_\mu}\left(F_{r^i}\frac{\partial\Gamma^i}{\partial\beta_\nu}\right) - \frac{\partial}{\partial\beta_\nu}\left(F_{r^i}\frac{\partial\Gamma^i}{\partial\beta_\mu}\right) \equiv 0, \qquad ((s,\beta) \in \Pi)$$

where F_{r^i} is evaluated as in (17.6)''.

The Hilbert integral is independent of admissible paths χ joining two points in $\Gamma(\Pi)$, if and only if the integral $H^*(\lambda)$ has a similar property for paths λ in Π. Since $\Gamma(\Pi)$ is simply-connected the latter condition is satisfied if and only if

(17.8) $$\frac{\partial M}{\partial\beta_\mu} \equiv \frac{\partial N_\mu}{\partial s}$$

(17.9) $$\frac{\partial N_\mu}{\partial\beta_\nu} \equiv \frac{\partial N_\nu}{\partial\beta_\mu}$$

$$\left\{\begin{array}{l}((s,\beta) \in \Pi)\\ (\nu,\mu = 1,\ldots,m)\end{array}\right\}$$

With the aid of (17.6)'' one sees that conditions (17.9) reduce to the conditions (17.7).

Turning to the conditions (17.8), let a superscript $*$ on F or on a partial derivative of F indicate evaluation with $(u,r) = (\Gamma(s,\beta),\Gamma_s(s,\beta))$ and $(s,\beta) \in \Pi$. Conditions (17.8) are equivalent to the identities

* For the present, paths will be regarded as admissible if piecewise regular.

$$\frac{\partial F^*}{\partial \beta_\mu} \equiv \frac{\partial}{\partial s}\left(F^*_{r^i}\Gamma^i_{\beta_\mu}\right) \qquad (\mu = 1, \ldots, m)(\alpha, \beta \in \Pi)$$

A brief computation verifies these identities on making use of the fact that the Euler equations $\frac{\partial}{\partial s} F^*_{r^i} = F^*_{u^i}$ hold for $i = 1, \ldots, n$ and for each β fixed in B.

Thus Theorem 17.1 is true.

The following lemma will be useful in applying Theorem 17.1.

Lemma 17.1. *Let $L_{\mu\nu}(s,\beta)$ denote the left member of (17.7), indexed by the integers μ, ν. Regardless of whether or not the family Γ is a field, $L_{\mu\nu}(s,\beta)$ is independent of the value of s for $(s,\beta) \in \Pi$.*

To establish this lemma we introduce the integral

$$J(s, \beta) = \int_{s_0}^{s} F(\Gamma(t, \beta), \Gamma_s(t, \beta))\, dt \qquad ((s, \beta) \in \Pi) \tag{17.10}$$

From (17.10) one finds that for $\mu, \nu = 1, \ldots, m$,

$$\frac{\partial J}{\partial \beta_\mu} \equiv \left[F^*_{r^i} \frac{\partial \Gamma^i}{\partial \beta_\mu}\right]_{(s_0, \beta)}^{(s, \beta)}, \quad \frac{\partial J}{\partial \beta_\nu} \equiv \left[F^*_{r^i} \frac{\partial \Gamma^i}{\partial \beta_\nu}\right]_{(s_0, \beta)}^{(s, \beta)} \tag{17.11)'}$$

From (17.11)' and the fact that

$$\frac{\partial^2 J}{\partial \beta_\mu \partial \beta_\nu} = \frac{\partial^2 J}{\partial \beta_\nu \partial \beta_\mu}$$

we infer that

(17.11)''

$$\left[\frac{\partial}{\partial\beta_\mu}\left(F^*_{r^i}\frac{\partial\Gamma^i}{\partial\beta_\nu}\right) - \frac{\partial}{\partial\beta_\nu}\left(F^*_{r^i}\frac{\partial\Gamma^i}{\partial\beta_\mu}\right)\right]^{(s,\beta)} = \left[\frac{\partial}{\partial\beta_\mu}\left(F^*_{r^i}\frac{\partial\Gamma^i}{\partial\beta_\nu}\right) - \frac{\partial}{\partial\beta_\nu}\left(F^*_{r^i}\frac{\partial\Gamma^i}{\partial\beta_\mu}\right)\right]^{(s_0,\beta)}$$

or equivalently $L_{\mu\nu}(s,\beta) \equiv L_{\mu\nu}(s_0,\beta)$.

Thus Lemma 17.1 is true.

Definition 17.1 prepares for Lemma 17.2.

Definition 17.1. A Mayer field. A field Γ on which the associated Hilbert integral is independent of admissible paths joining two points in the field, is called a Mayer field.

The following lemma is an aid in proving the principal theorem of §18.

Lemma 17.2. *Let $\underset{\sim}{z}$: (17.0) be an extremal arc of a Weierstrass integral J_F in the domain U of a presentation (ϕ, U). If the initial point of $\underset{\sim}{z}$ is conjugate to no point on $\underset{\sim}{z}$ there exists a Mayer field Γ_0 on U such that the extremal arc $\underset{\sim}{z}$ is a proper subarc of an extremal of Γ_0 of form*

(17.12) $$s \to \gamma(s) : [s_0, s_1] \to U$$

where $\beta = \underset{\sim}{0}$ *and* $s_0 < a < b < s_1$.

An extension $\underset{\sim}{z}^E$ of $\underset{\sim}{z}$. The extremal arc $\underset{\sim}{z}$ admits an extension $\underset{\sim}{z}^E$ in U which is a compact extremal arc on which $\underset{\sim}{z}$ is a proper subarc. We suppose that $\underset{\sim}{z}^E$ is parameterized by R-length s, measured from the initial point of $\underset{\sim}{z}^E$, so that on $\underset{\sim}{z}^E$, $0 \le s \le d$, where d is the R-length of $\underset{\sim}{z}^E$. As a proper subarc of $\underset{\sim}{z}^E$, $\underset{\sim}{z}$ will be represented by a subinterval $[a, b]$ of values of s such that $0 < a < b < d$. We shall apply Theorem 10.3 on the continuous variation of the first conjugate point of a point $s = c$ with c.

A choice of c, s_0 and s_1. Given a and b, let c and s_1 be values in R such that

(17.13)' $$0 < c < a < b < s_1 < d .$$

By hypothesis there is no conjugate point of a on the interval $(a, b]$. It follows from Theorem 10.3 that if the interval $[c, s_1]$ differs sufficiently little from the interval $[a, b]$ there will be no conjugate point of c on the interval $(c, s_1]$. We suppose $[c, s_1]$ so chosen and choose s_0 so that

(17.13)'' $$0 < c < s_0 < a < b < s_1 < d.$$

Definition of a Mayer field Γ_0. According to Lemma 9.2 there exists a proper polar family $\underset{\sim}{\Phi}$ of extremal arcs of J_F with $\underset{\sim}{z}^E$ as central extremal and $s = c$ its pole. $\underset{\sim}{\Phi}$ is defined as in §9 by a C^∞-mapping,

(17.14) $$(s, \beta) \to \Phi(s, \beta) : [0, d] \times B_m \to U .$$

Let B^ρ be an origin-centered open m-ball of radius ρ included in B_m. Set

(17.15) $$\Phi | ([s_0, s_1] \times B^\rho) = \Gamma_0 .$$

If ρ is sufficiently small, the choice of c, s_0, s_1 is such that the family of extremal arcs

$$(17.16) \qquad s \to \Gamma_0(s,\beta) : [s_0, s_1] \to U \qquad (\beta \in B^\rho)$$

will be a field Γ_0 in U. The extremal $\underset{\sim}{z}$ is a <u>proper</u> subarc of Γ_0^β when $\beta = \underset{\sim}{0}$. The field Γ_0 will cover the subset $\Gamma_0(\Pi_0)$ of U, where $\Pi_0 = [s_0, s_1] \times B^\rho$. It remains to prove the following.

(i) <u>The field Γ_0, defined by (17.15), is a Mayer field of extremals.</u>

Let $L_{\mu\nu}(s,\beta)$ be the left member of (17.7) obtained by replacing Γ in (17.7) by Φ for $[s,\beta]$ in $[0,d] \times B_m$, the domain of Φ. Since $\underset{\sim}{\Phi}$ has a pole when $s = c$, $L_{\mu\nu}(c,\beta) \equiv 0$ for $\beta \in B_m$ for each pair μ,ν. It follows from Lemma 17.1 that $L_{\mu\nu}(s,\beta) \equiv 0$ for (s,β) in the domain of $\underset{\sim}{\Phi}$. In particular

$$(17.17) \qquad L_{\mu\nu}(s,\beta) \equiv 0 \qquad ((s,\beta) \in \Pi_0,\ \mu,\nu = 1,\ldots,m)$$

Now the restrictions of $L_{\mu\nu}$ which are evaluated in (17.17) are given by the left members of (17.7) when Γ is replaced by Γ_0. According to Theorem 17.1, Γ_0 is then a Mayer field, as affirmed in (i).

The proof of Lemma 17.2 is complete.

§18. Three conditions sufficient for a relative minimum. In this section we continue with the extremal arc

$$\underset{\sim}{z} : s \to z(s) : [a, b] \to U \qquad \text{(cf. (17.0))} \tag{18.1}$$

of a W-integral J_F associated with a presentation $(\phi, U) \in \mathcal{D}M_n$. Given an open neighborhood N of $|\underset{\sim}{z}|$ in U, piecewise regular curves $\underset{\sim}{u}$ joining the endpoints of $\underset{\sim}{z}$ in N will be termed admissible in N. The integrals $J_F(\underset{\sim}{z})$ and $J_F(\underset{\sim}{u})$ will be compared. A minimum, afforded to J_F by $\underset{\sim}{z}$, will be termed a strong minimum, relative to N, if there are no conditions on the curves $\underset{\sim}{u}$ joining the endpoints of $\underset{\sim}{z}$ in N other than that the curves $\underset{\sim}{u}$ be piecewise regular. A minimum so afforded to J_F is termed proper if $J_F(\underset{\sim}{z}) = J_F(\underset{\sim}{u})$ for no admissible R-parameterized curves $\underset{\sim}{u}$ in N other than $\underset{\sim}{z}$.

The following three conditions will be shown to imply that the extremal arc $\underset{\sim}{z}$ afford a proper, strong, relative minimum to J_F.

I. The W-nonsingularity of F. This is the condition imposed at the end of §6 on each preintegrand F. It insures the validity of the theorems of §7 on solutions of the Euler-Riemann equations.

II. The positive regularity of F. This is the condition that F be positive regular in the sense of Definition 14.2. According to Theorems 14.3 and 15.1, this condition is necessary if F is nonsingular and if $\underset{\sim}{z}$ affords a weak relative minimum to J_F.

III. <u>The Jacobi* S-condition on $\underset{\sim}{z}$.</u> This is the condition that the initial point p of $\underset{\sim}{z}$ be conjugate to no point on $\underset{\sim}{z}$. This condition is not necessary, as simple examples will show. However, Theorem 16.1 shows how the Jacobi S-condition can be relaxed in order to be necessary.

A lemma that prepares for Theorem 18.1 follows.

<u>Lemma</u> 18.1. <u>Under Conditions</u> I <u>and</u> II <u>on the preintegrand</u> F

$$\mathcal{E}_F(u, \rho, r) > 0 \tag{18.2}$$

<u>for</u> $u \in U$ <u>and arbitrary unequal n-tuples,</u> ρ <u>and</u> r <u>which are</u> R-<u>unitary</u>[†] <u>at</u> u.

Under the conditions on ρ and r of Lemma 18.1 just two cases arise:

<u>Case</u> I. $r \neq -\rho$.   <u>Case</u> II. $r = -\rho$.

<u>Proof in Case</u> I. Under the conditions of the lemma, ρ and r are linearly dependent only if $r = -\rho$. In Case I the formula (5.22) for $\mathcal{E}_F(u, \rho, r)$ accordingly applies. By hypothesis F is positive regular, so that the right member of (5.22) is positive, unless for some scalar k

$$r - \rho = kX = k(\rho + \theta(r - \rho)) \qquad (0 < \theta < 1)$$

or equivalently

$$r(1 - k\theta) = \rho(k + 1 - k\theta) \,. \tag{18.3}$$

Since r and ρ are linearly independent in Case I, (18.3) is possible only if

---

*S abbreviates sufficient.

[†]An n-tuple r is R-<u>unitary</u> at u if $\mathcal{f}(u, r) = 1$. See (2.6) for $\mathcal{f}(u, r)$.

$1 - k\theta$ and $k + 1 - k\theta$ both vanish, which is impossible. Hence in Case I the right member of (5.22) is positive and (18.2) is true.

Proof in Case II. Let σ be an n-tuple which is R-unitary at $u \in U$, but which equals neither r nor $-r$. The definition of $\mathcal{E}_F$ in (5.16) shows that

$$\mathcal{E}_F(u, r, -r) = \mathcal{E}_F(u, \sigma, -r) + \mathcal{E}_F(u, \sigma, r) \tag{18.4}$$

The validity of Lemma 18.1 in Case I implies that both terms on the right of (18.4) are positive. Hence $\mathcal{E}_F(u, -r, r) > 0$, so that (18.2) is true in Case II.

This completes the proof of Lemma 18.1.

Amended Lemma 18.1. If the word "unequal" is deleted from Lemma 18.1, Lemma 18.1, amended to read $\mathcal{E}_F(u, \rho, r) \geq 0$, is true.

The principal theorem of this section follows.

Theorem 18.1. *If the preintegrand** F *is positive regular and if the Jacobi S-condition is satisfied on the extremal* $\underset{\sim}{z}$ *of* J_F, *then* $\underset{\sim}{z}$ *affords a proper, strong minimum to* J_F *relative to piecewise regular curves* $\underset{\sim}{h}$ *which join the endpoints of* $\underset{\sim}{z}$ *in a sufficiently small open neighborhood* N *of* $|\underset{\sim}{z}|$ *in* U.

In defining the Hilbert integral (17.4) use was made of a field Γ. In proving Theorem 18.1 we identify Γ with the Mayer field Γ_0 affirmed to exist in Lemma 17.2. Except for the proof of Lemma 17.2 the proof of Theorem 18.1 follows the classical Weierstrass model.

---

*The nonsingularity of preintegrands F has been assumed since the end of §6.

Any piecewise R-parameterized regular curve

$$(18.5)\qquad \underset{\sim}{h} : s \to h(s) : [a, b_h] \to U$$

is admitted which joins the endpoints of $\underset{\sim}{z}$ in the carrier $\Gamma(\Pi)$ of the field $\Gamma = \Gamma_0$. We refer to the Hilbert integral H_F over curves $\underset{\sim}{h}$ and begin with the proof of three formulas:

$$(18.6)\qquad H_F(\underset{\sim}{z}) = J_F(\underset{\sim}{z})$$

$$(18.7)\qquad H_F(\underset{\sim}{z}) = H_F(\underset{\sim}{h}) = \int_{\underset{\sim}{h}} F_{r^i}(u, \rho(u))\, du^i$$

$$(18.8)\qquad J_F(\underset{\sim}{h}) - J_F(\underset{\sim}{z}) = \int_a^{b_h} \mathcal{E}_F\big(h(s), \rho(h(s)), \dot{h}(s)\big)\, ds$$

<u>Verification of</u> (18.6). The Hilbert integral $H_F(\underset{\sim}{z})$ takes the form (17.4), with $\Gamma = \Gamma_0$ in (17.3), and $\underset{\sim}{z}$ replacing χ, and can be evaluated as a line integral (17.5). So evaluated $H_F(\underset{\sim}{z})$ reduces to $J_F(\underset{\sim}{z})$, since along $\underset{\sim}{z}$ the differentials $d\beta_\mu = 0$ by virtue of Lemma 17.2.

<u>Verification of</u> (18.7). The independence of path of the Hilbert integral along paths joining two points in the Mayer field $\Gamma = \Gamma_0$ implies (18.7).

<u>Verification of</u> (18.8). By virtue of (18.6) and (18.7)

$$(18.9)\quad J_F(\underset{\sim}{h}) - J_F(\underset{\sim}{z}) = J_F(\underset{\sim}{h}) - H_F(\underset{\sim}{z}) = \int_a^{b_h} [F(h(s), \dot{h}(s)) - F_{r^i}\big(h(s), \rho(h(s))\big)\dot{h}^i(s)]\, ds$$

thereby establishing (18.8).

Completion of proof of Theorem 18.1. According to the amended Lemma 18.1 the values of the integrand $\mathcal{E}_F$ of the integral on the right of (18.8) are nonnegative. Theorem 18.1 follows if the word "proper" is deleted.

The integral on the right of (18.8) will vanish only if its integrand is zero for each value of s for which $\dot{h}(s)$ is defined, or equivalently, according to Lemma 18.1, only if at each point u on each regular subarc $\underset{\sim}{u}$ $s \to u(s)$ of $\underset{\sim}{h}$, the R-unitary direction

$$\frac{du}{ds} = \rho(u) \tag{18.10}$$

The DE (18.10) defines the extremal arcs of the field Γ and is accordingly satisfied by the arc $s \to u(s)$ only if $\underset{\sim}{u}$ is an extremal arc of the field Γ. Since this is true for each regular subarc of $\underset{\sim}{h}$, $\underset{\sim}{h}$ must be identical with $\underset{\sim}{z}$.

This completes the proof of Theorem 18.1.

A second sufficient condition for positive regularity of W-preintegrands F. A nonsingular W-preintegrand associated with a presentation (ϕ, U) is not necessarily positive regular. This is clear: when F is Pos-R, -F is not Pos-R; while -F remains nonsingular if F is nonsingular. The following gives a simple condition that a nonsingular F be Pos-R. When $n > 2$, Theorem 18.2 is believed new.

Theorem 18.2. Let a presentation $(\phi, U) \in \mathcal{E}M_n$ be given and an associated nonsingular preintegrand F. If for some point $u_0 \in U$ and every

<u>nonnull</u> n-<u>tuple</u> r, $F(u_0, r) > 0$, <u>then</u> F <u>is positive regular in the sense of Definition</u> 14.2.

Because of the invariance of positive regularity under a transition diff we can suppose u_0 is the origin in U. Let ρ be an n-tuple such that

$$(18.11)' \qquad F(\underset{\sim}{0}, \rho) = \min F(\underset{\sim}{0}, r) \mid (\|r\| = 1)$$

We can suppose that

$$(18.11)'' \qquad \rho = (1, 0, \ldots, 0)$$

Were (18.11)'' not valid one could subject the coordinate domain to a transition diff θ and the n-tuples r to the corresponding contravariant transformation, supposing that θ maps the origin $u_0 = \underset{\sim}{0}$ into the origin. One could, in particular, take θ as an orthogonal transformation T of the n-tuples u. The corresponding contravariant transformation of the n-tuples r would be the same orthogonal transformation. T could be chosen so that the n-tuple ρ for which (18.11)' holds, is carried by T into the n-tuple (1, 0, . . . , 0).

We can accordingly suppose that the n-tuple ρ for which (18.11)' holds is the n-tuple (18.11)'' and that $u_0 = \underset{\sim}{0}$. Let Π be the m-plane in R^n on which $r^1 = 1$ and let S be the (n-1)-sphere in R^n on which $\|r\| = 1$. Π is tangent to S at the point $\rho = (1, 0, \ldots, 0)$. We can then verify the following:

(i) <u>For</u> $\sigma \in \Pi$, $F(\underset{\sim}{0}, \sigma) \geq F(0, \rho)$;

(ii) <u>The quadratic form with</u> (n-1)-<u>square matrix</u>

$$\| F_{r^i r^j}(\underset{\sim}{0}, \rho) \| \qquad (i, j = 2, 3, \ldots, r) \tag{18.12}$$

is positive definite;

(iii) F is positive regular at $(\underset{\sim}{0}, \rho)$.

Proof of (i). According to the definition (18.11)' of ρ, $F(\underset{\sim}{0}, r) \geq F(\underset{\sim}{0}, \rho)$ for $\|r\| = 1$. Corresponding to each n-tuple $\sigma \in \Pi$ there exists an n-tuple $r_\sigma \in S$ and a scalar $k_\sigma \geq 1$ such that the product $k_\sigma r_\sigma = \sigma$. Hence for $\sigma \in \Pi$

$$F(\underset{\sim}{0}, \sigma) = k_\sigma F(\underset{\sim}{0}, r_\sigma) \geq F(\underset{\sim}{0}, \rho) \tag{18.13}$$

thereby verifying (i).

Proof of (ii). Set $a_{ij} = F_{r^i r^j}(\underset{\sim}{0}, \rho)$ for $(i, j) = 1, \ldots, n$. Let A^{11} be the cofactor of a_{11} in the determinant $|a_{ij}|$. It follows from (6.9) that $A^{11} \neq 0$, since the Weierstrass coefficient $F_1(\underset{\sim}{0}, \rho) \neq 0$. It follows from (i) that the quadratic form with matrix (18.12) is positive semi-definite. The determinant of the matrix (18.12) is $A^{11} \neq 0$. The quadratic form with matrix (18.12) must then be positive definite.

Proof of (iii). That each element in the first row and column of $\|a_{ij}\|$ vanishes, follows from (4.4) with $r = (1, 0, \ldots, 0)$ therein. With this understood, (iii) follows from (ii).

That F is positive regular at each 2n-tuple (u, r) in its domain follows from Theorem 15.1.

Thus Theorem 18.2 is true.

Corollary 18.1. *Let a presentation* (ϕ, U) *be given and an associated nonsingular preintegrand* F. *If* F *is positive definite then* F *is positive regular.*

In particular, each R-preintegrand f is positive regular. For f is nonsingular by Theorem 6.2. It is clearly positive definite and hence, by Corollary 18.1, it is positive regular. Theorem 18.1 accordingly yields the following classical theorem concerning R-length on a Riemannian manifold M_n.

Corollary 18.2. *A geodesic arc* $\mathbf{z}$ *of* M_n *in a coordinate range* $\phi(U)$ *on* M_n *whose initial point is conjugate to no point on* $\mathbf{z}$ *including the terminal point of* $\mathbf{z}$, *affords a proper, strong, minimum to* R-*length on* M_n *relative to any piecewise regular curve on* M_n *which joins the endpoints of* $\mathbf{z}$ *in a sufficiently small open neighborhood* N *of* $|\mathbf{z}|$ *in* $\phi(U)$.

The converse of Corollary 18.1 is false. That is, if F is positive regular it does not follow that F is positive definite. The following example shows this.

Example 18.1. When $n = 2$ let

$$F(u, r) = -2r^1 + \|r\| .$$

Such an F is positively homogeneous in the 2-tuple r. It is positive regular and hence nonsingular. For the 2-tuples $\rho = (1, 0)$ and $\sigma = (0, 1)$, $F(u, \rho) = -1$, $F(u, \sigma) = 1$, respectively.

PART IV

Preparation for Global Theorems

Chapter 7

Elementary Extremals

§19. <u>Field radii and elementary extremals</u>. We begin with an example. In the $[u^1, u^2]$-plane the extremals of the integral of length are straight lines. The extremals issuing from the origin with lengths at most 1 have a representation

$$[u^1, u^2] = [s \cos \beta, \ s \sin \beta] \qquad [0 \leq s \leq 1] \tag{19.1}$$

with β a parameter. These extremal arcs define a "field" when $s > 0$, with the origin as "pole" when $s = 0$. We say that this field has a <u>field radius</u> 1. The parameters β represent points $[r^1, r^2]$ on the circle

$$C : (r^1)^2 + (r^2)^2 = 1. \tag{19.2}$$

This field is given by a mapping of class C^∞ of the product $[0,1] \times C$ into R^2. The mapping is biunique when $s > 0$.

<u>Objectives and hypotheses</u>. We are concerned with the representation of extremals of J issuing from a point $p \in M_n$, with R-lengths restricted by a suitable positive constant. The point p is termed the <u>pole</u>. Three aspects of such representations require study.

(A_1) <u>The differentiability of the representations</u>.

(A_2) <u>Biuniqueness, the pole excepted</u>.

(A_3) <u>The global minimizing properties of the extremals</u>.

It will be sufficient to study the extremals of an integral J_F associated with an arbitrary presentation $(\phi, U) \in \mathcal{D}M_n$. A general hypothesis that M_n be compact is made. The special hypotheses to be made under (A_1), (A_2), (A_3) are, respectively:

(a_1) The nonsingularity of preintegrands F.

(a_2) The nonsingularity and positive regularity of preintegrands F.

(a_3) The nonsingularity and positive definiteness of preintegrands F.

The hypothesis (a_1) is insufficient for (A_2) while hypotheses (a_1) and (a_2) combined are insufficient for (A_3). There is here a striking difference between the study of W-integrals and the simpler study of R-integrals. See Example 18.1.

(A_1) <u>The differentiability of the extremal representations.</u> There is given a presentation $(\phi, U) \in \mathcal{D}M_n$ with its associated W-preintegrand F and R-preintegrand $\mathcal{f}$. We suppose that $\mathcal{f}$ has values

$$\mathcal{f}(u, r) = (a_{ij}(u)r^i r^j)^{\frac{1}{2}} \qquad ((u, r) \in U \times \dot{R}^n) \tag{19.3}$$

A point $u_0 \in U$ and image $\phi(u_0) = p_0 \in M_n$ are given. The analogue of the circle C : (19.2) is the ellipsoid $\mathcal{f}^{u_0}$ defined by the condition,

$$1 = (a_{ij}(u_0)r^i r^j)^{\frac{1}{2}} = \mathcal{f}(u_0, r) . \tag{19.4}$$

The n-tuples r on $\mathcal{f}^{u_0}$ are called <u>unitary</u>. See Definition 7.3. It is fundamental that $\mathcal{f}^{u_0}$ is a <u>compact</u> subset of R^n.

Lemmas 19.1 and 19.2 prepare for Theorem 19.1.

<u>The Differentiability Lemma</u> 19.1. <u>Corresponding to a presentation $(\phi, U) \in \mathcal{D}M_n$ with an associated nonsingular</u> W-<u>preintegrand</u> F <u>and</u> R-<u>preintegrand</u> $\mathcal{f}$: (19.3), <u>and to a point</u>* $u_0 \in U$, <u>there exists a positive</u>

* For simplicity we suppose that u_0 is the origin.

constant L_0 such that the following is true.

Corresponding to each n-tuple $r \in \mathcal{J}^{u_0}$, there exists an extremal

(19.5) $$s \to \omega(s, r) : [0, L_0] \to U$$

of J_F with initial point the origin $u_0 = \omega(0, r)$ and initial unitary n-tuple $r = \omega_s(0, r)$ prescribed in $\mathcal{J}^{u_0}$. The mapping

(19.5)' $$(s, r) \to \omega(s, r) : [0, L_0] \times \mathcal{J}^{u_0} \to U ,$$

thereby defined, is of class C^∞.

This lemma is a consequence of Theorem 7.1. Theorem 7.1 represents the RL-solutions of the Euler-Riemann equations (7.0) with initial points a and initial t-derivatives b. The 2n-tuple (a, b) is in an open neighborhood of the initial 2n-tuple (u_0, r_0). Here, however, one fixes a as the origin u_0 and restricts the n-tuple r to a sufficiently small neighborhood, relative to $\mathcal{J}^{u_0}$, of $r_0 \in \mathcal{J}^{u_0}$. Lemma 19.1 follows, on recalling that $\mathcal{J}^{u_0}$ is compact.

(A_2) Biuniqueness, the origin excepted. The following lemma is proved under the assumption that F is nonsingular and positive regular. F need not be positive definite in this lemma.

Lemma 19.2. If $k < L_0$ is a sufficiently small positive constant the subarcs

(19.6) $$s \to \omega(s, r) : [0, k] \to U \qquad [r \in \mathcal{J}^{u_0}, u_0 = \underset{\sim}{0}]$$

of the extremals (19.5) are

(i) conjugate point free, and

(ii) nonintersecting when $s > 0$.

Proof of (i). Let $\underset{\sim}{z}$ be an extremal arc of form (19.6) with initial n-tuple $r_0 \in \mathcal{L}^{u_0}$. If $k > 0$ is sufficiently small there will be no conjugate point of $s = 0$ on $\underset{\sim}{z}$ when $0 < s \leq k$. It follows that this is also true for all extremals (19.6) for which $r \in \mathcal{L}^{u_0}$ and $\|r - r_0\|$ is sufficiently small. Since $\mathcal{L}^{u_0}$ is compact, (i) is still true if the condition on $\|r - r_0\|$ is dropped, provided $k > 0$ is sufficiently small.

Proof of (ii). Suppose (ii) false. That is, suppose that there exists a decreasing sequence

$$c_1 > c_2 > c_3 > \cdots$$

of positive constants $c_\mu < L_0$ which converge to 0 as $\mu \uparrow \infty$ and for which the following is true. For each integer μ there exist extremals $\omega(\cdot, r_\mu)$ and $\omega(\cdot, \sigma_\mu)$ of the family (19.5) with different initial unitary n-tuples $r \in \mathcal{L}^{u_0}$ such that for values s_μ and t_μ in the interval $(0, c_\mu]$

$$\omega(s_\mu, r_\mu) = \omega(t_\mu, \sigma_\mu). \tag{19.7}$$

We can replace the sequence (c_μ) by a subsequence such that as $\mu \uparrow \infty$, r_μ and σ_μ tend, respectively, to limiting n-tuples $\bar{r}$ and $\bar{\sigma}$ on $\mathcal{L}^{\mu_0}$.

We are supposing that for each μ

$$\omega(0, r_\mu) = \omega(0, \sigma_\mu) = 0; \; \omega_s(0, r_\mu) = r_\mu; \; \omega_s(0, \sigma_\mu) = \sigma_\mu \tag{19.8}$$

From Taylor's formula applied to the n-tuple $\omega(s, r) - \omega(0, r)$ for a fixed r we find that

$$\omega(s_\mu, r_\mu) = \omega_s(0, r_\mu)s_\mu + A(s_\mu, r_\mu)s_\mu^2 \tag{19.9)'}$$

$$\omega(t_\mu, \sigma_\mu) = \omega_s(0, \sigma_\mu)t_\mu + B(t_\mu, \sigma_\mu)t_\mu^2 \tag{19.9)''}$$

where $A(s_\mu, r_\mu)$ and $B(t_\mu, \sigma_\mu)$ are bounded independently of the choice of μ. The relations (19.7) to (19.9) imply that for each μ

$$s_\mu[r_\mu + s_\mu A(s_\mu, r_\mu)] = t_\mu[\sigma_\mu + t_\mu B(t_\mu, r_\mu)] \qquad (s_\mu > 0,\ t_\mu > 0) . \tag{19.10}$$

The brackets in (19.10) represent n-tuples, which converge respectively to $\bar{r}$ and $\bar{\sigma}$ as $\mu \uparrow \infty$. We infer the following:

(α) The relations (19.10) hold for $\mu = 1, 2, \ldots,$ only if $\bar{r} = \bar{\sigma}$.

Let $\underset{\sim}{w}$ be an initial subarc of the extremal $\omega(\cdot, \bar{r}) = \omega(\cdot, \bar{\sigma})$, taken so short that there is no conjugate point of $s = 0$ on $\underset{\sim}{w}$. Concerning $\underset{\sim}{w}$ the following can be affirmed. The terminology is that of §14.

(β) Corresponding to each positive constant e the e-tube T of Ph $\underset{\sim}{w}$ carries the phase images of a pair of extremals of J_F which intersect in $u_0 = \underset{\sim}{0}$, and in at least one other point in U. According to Corollary 1 of Appendix II this is impossible since $\underset{\sim}{w}$ is conjugate point free. Thus (ii) is true.

The proof of Lemma 19.2 is complete.

Lemma 19.2 has the following uniform extension.

Theorem 19.1. Suppose that M_n is compact and that each W-preintegrand is nonsingular and positive regular.

If then κ is a sufficiently small positive constant, extremal arcs of J with R-lengths at most κ issuing from an arbitrary point $p \in M_n$ will be such that

(i) each extremal g is conjugate point free,

(ii) extremals issuing from p intersect only in p.

Each point p is contained in the range X_p of a presentation in $\mathcal{D}M_n$. Difficulties will arise in the proof of Theorem 19.1 if the boundaries of the ranges are not uniformly bounded from the respective points p. For this reason the reader is asked to confirm the following statement (A).

(A) There exists a positive constant L_0 so small that the following is true. Corresponding to each point $p \in M_n$ there exists a presentation (ϕ_p, U_p) in $\mathcal{D}M_n$ such that U_p contains the origin $\underset{\sim}{0} \in R^n$, $\phi(\underset{\sim}{0}) = p$ and $\phi_p(U_p)$ contains each point in M_n whose R-distance from p on M_n is less than L_0. We term the presentations (ϕ_p, U_p) canonical with pole p.

Proof of (i). The assumption that (i) is false will be shown to be contrary to Theorem 1 of Appendix II.

If (i) is false there exists a sequence (p_μ) of points in M_n such that Lemma 19.2 holds when (ϕ, U) is replaced by $(\phi_{p_\mu}, U_{p_\mu})$, but only if k in Lemma 19.2 is a value $k_\mu < \frac{1}{\mu}$. We can suppose the points p_μ chosen so that they converge as $\mu \uparrow \infty$ to a point $q \in M_n$.

Let (ϕ_q, U_q) be the canonical presentation with pole q. See (A). Let F_q be the W-preintegrand associated with (ϕ_q, U_q). From Lemma 19.2 we infer the following. There exists a positive constant κ_q such that extremals of J_{F_q} issuing from the origin $\underset{\sim}{0}$ in U_q bear no conjugate point of $\underset{\sim}{0}$ when $0 < s < \kappa_q$.

Because $k_\mu < \frac{1}{\mu}$ for each positive integer μ there will issue from $q \in U_q$ an extremal $\underset{\sim}{w}$ of J_{F_q} of R-length κ_q which is conjugate point free, but in every neighborhood of which lie extremal arcs of J_{F_q} which are not conjugate point free but which are phasewise arbitrarily near $\underset{\sim}{w}$. This is impossible by Theorem 10.6 or Theorem 1 of Appendix II. Thus (i) must be true.

Proof of (ii). Were (ii) false one could show, as in the proof of (i), that Corollary 1 of Appendix II is contradicted.

We conclude that Theorem 19.1 is true.

Note. The mapping (19.5)' of Lemma 19.1 is onto a neighborhood of the origin $\underset{\sim}{0}$ in U. If S^ρ_{n-1} is any origin-centered (n-1)-sphere in U of sufficiently small radius ρ, each point of S^ρ_{n-1} is met by an extremal of J_F issuing from 0. Statement (i) of Lemma 19.2 is needed to prove this.

(A_3) The global minimizing properties of elementary extremals. In the second and final theorem of this section we shall assume that each preintegrand F is nonsingular and positive definite. The hypotheses of Theorem 19.1 do not imply that F is positive definite. In preparation for Theorem 19.2 several definitions are required.

Definition 19.1. J-lengths. When each F is positive-definite the value of J on an admissible curve will be called its J-length.

Definition 19.2. Field radii. A positive constant $\kappa < L_0$ satisfying Theorem 19.1 will be called a field radius of J. L_0 is the constant of $(\underset{\sim}{A})$.

It is clear that if κ is a field radius, any positive constant less than κ also satisfies Theorem 19.1. More generally, if κ is a field radius there exists a positive constant ε so small that each number k such that $0 < k < \kappa+\varepsilon$ is also a field radius.

Definition 19.3. The J-bounds $m^p(\kappa)$ and $m(\kappa)$. Let κ be a field radius. Fix κ. Corresponding to a point $p \in M_n$ let $m^p(\kappa)$ be the minimum of the J-lengths of extremals of J which issue from p and have R-lengths κ. For fixed κ set

$$m(\kappa) = \underset{p \in M_n}{\text{minimum}}\ m^p(\kappa) . \tag{19.11}$$

On making use of the coordinate domain of a canonical presentation (ϕ_p, U_p) with pole p one shows readily that for fixed κ the mapping $p \rightarrow m^p(\kappa)$ is continuous at each point $p_0 \in M_n$. Since M_n is compact, the minimum $m(\kappa)$ exists and is positive.

Definition 19.4. Elementary extremals of J. Of the R-lengths κ which are field radii we shall prefer one and denote it by $\underset{\sim}{\kappa}$. The corresponding J-length $m(\kappa)$ of Definition 19.3 will then be denoted by $\underset{\sim}{m}$. Any extremal ξ of J whose J-length is at most $\underset{\sim}{m}$ will be called an elementary extremal of J. The numbers $\underset{\sim}{\kappa}$ and $\underset{\sim}{m}$ will be called the preferred field radius and

J-length.

If (ϕ, U) is a presentation in $\mathcal{D}M_n$ and F an associated W-preintegrand, respectively, an extremal ζ of J_F such that

$$J_F(\zeta) \leq \underset{\sim}{m}, \tag{19.12}$$

will be called an elementary extremal of J_F.

We state a basic theorem.

Theorem 19.2. If M_n is compact and each preintegrand F, non-singular and positive definite, the following is true:

(i) If ξ is an arbitrary elementary extremal of J, then

$$J(\xi) \leq J(\gamma) \tag{19.13}$$

where γ is any piecewise regular curve which joins the endpoints of ξ on M_n.

(ii) The equality holds in (19.13) only if the R-parameterization of γ is identical with that of ξ.

In proving Theorem 19.2 use will be made of the following topological n-discs.

Definition 19.5. Extremal n-discs D_ξ^p on M_n. Let ξ be an extremal of J with initial point p and a J-length which is at most $\underset{\sim}{m}$. The set of all extremals which issue from p and have J-lengths equal to $J(\xi)$ has a carrier on M_n which we denote by D_ξ^p and term an extremal n-disc on M_n with pole p. See Theorem 19.1.

Extremal n-discs D_ξ^p, as defined, can be represented in suitably chosen coordinate domains, as follows.

Definition 19.6. A local representative $\mathcal{L}_\zeta^F$ of an n-disc D_ξ^p. Corresponding to an extremal n-disc D_ξ^p there exists a presentation (ϕ, U) of the type characterized in $(\underset{\sim}{A})$ with $\phi(\underset{\sim}{0}) = p$ and with a preintegrand F such that the following is true. There exists an extremal ζ of J_F such that $\phi(\zeta) = \xi$, and a topological n-disc $\mathcal{D}_\zeta^F \subset U$ such that

$$\phi(\mathcal{L}_\zeta^F) = D_\xi^p .$$

The following lemma will imply Theorem 19.2(i).

Lemma 19.3. The elementary extremal ζ of F, representing ξ, as in Definition 19.6, is such that

$$J_F(\zeta) \leq J_F(h) \tag{19.14}$$

where h is any admissible curve joining the endpoints of ζ on $\mathcal{D}_\zeta^F$.

Lemma 19.3 can be simplified as follows. There exists a subcurve g of h such that g meets the origin $\underset{\sim}{0}$ and the boundary of $\mathcal{D}_\zeta^F$ only at g's initial endpoint $\underset{\sim}{0}$ and its final endpoint $\underset{\sim}{a}$. If η is the extremal of J_F which joins $\underset{\sim}{0}$ to $\underset{\sim}{a}$ then

$$J_F(\eta) \leq J_F(g) \tag{19.15}$$

as we shall prove. Since F is positive definite, (19.15) implies (19.14).

<u>Notation for the proof of</u> (19.15). It will simplify the proof of (19.15) if the value of J_F on a curve λ of U is denoted by $[\lambda]$. With this notation $[\eta] = [\zeta] = J(\xi)$. We shall prove (19.15) or equivalently that

$$[g] \geq [\eta] . \tag{19.16}$$

Admissible curves $\gamma_1, \ldots, \gamma_r$ will be defined to form a sequence of curves,

$$g, \gamma_1, \ldots, \gamma_r, \eta . \tag{19.17}$$

Each of these curves, excepting the extremal η, shall be in a small neighborhood of its successor and join the origin in U to the terminal point $\underset{\sim}{a}$ of ζ. The sequence shall be such that

$$[g] \geq [\gamma_1] \geq [\gamma_2] \geq \cdots \geq [\gamma_r] \geq [\eta] . \tag{19.18}$$

Once the curves $\gamma_1, \ldots, \gamma_r$ have been defined and (19.18) proved, (19.16) follows.

<u>Definition of the curves</u> $\gamma_1, \ldots, \gamma_r$. Suppose that g has an R-parameterization,

$$s \to g(s) : [0, b] \to U . \tag{19.19}$$

Let

$$0 < t_1 < t_2 < \cdots < t_r < b \tag{19.20}$$

be a set of values with equal successive differences.

The integer r shall be so large that a condition, presently to be defined, is satisfied. Let g_i denote the terminal subcurve of g on which $t_i \leq t \leq b$. Here $i = 1, \ldots, r$. Let E_i denote the extremal arc of J_F which joins $\underset{\sim}{0}$ to $g(t_i)$. The extremal arc E_i followed by g_i is our definition of γ_i. With this understood we shall write

$$\gamma_i = E_i * g_i . \tag{19.21}$$

<u>The proof of</u> (19.18). Set $t_0 = 0$ and $t_{r+1} = b$. For $k = 0, 1, \ldots, r$ let $\Delta_k g$ denote the subcurve of g on which $t_k \leq s \leq t_{k+1}$. Set $\eta = \gamma_{r+1}$. The curve γ_k is obtained from the curve γ_{k+1} by replacing the extremal subcurve E_{k+1} of γ_{k+1} by the curve $E_k * \Delta_k g$ defined by tracing E_k and then $\Delta_k g$. The curve $E_k * \Delta_k g$ joins the endpoints of E_{k+1}. To establish (19.18) one has merely to show that the J-length

$$[E_{k+1}] \leq [E_k * \Delta_k g] \qquad (k = 0, 1, \ldots, r) \tag{19.22}$$

if the integer r is sufficiently large.

<u>Proof of</u> (19.22). A proof of (19.22) involves a question of uniformity which needs to be clarified. Given $\mathcal{L}^G_\zeta$ let $\underset{\sim}{E}$ denote the class of extremals E of J_F which join the origin to the boundary of $\mathcal{L}^F_\zeta$. We affirm (α).

(α) <u>If</u> e <u>is a sufficiently small positive constant and</u> N^e_E <u>the e-neighborhood in</u> U <u>of an arbitrary extremal</u> $E \in \underset{\sim}{E}$, <u>then the</u> J_F<u>-length</u> $[E]$ <u>is less than</u>

<u>or equal to the</u> J_F<u>-length</u> $[\lambda]$ <u>of any admissible curve which joins the endpoints of</u> E <u>in</u> N_E^e.

A proof of statement (a) is an easy consequence of Theorem 1 of Appendix II. With (a) understood, let the integer r of (19.18) be so large that the curve $(E_k * \Delta_k g)$ of (19.22) is in the e-neighborhood of E_{k+1}. It follows from statement (a) that (19.22) is true.

For such a choice of r, (19.18) is true and hence Lemma 19.3.

<u>Completion of proof of Theorem</u> 19.2. Since each preintegrand F is positive definite, Lemma 19.3 implies (i) of Theorem 19.2.

Statement (ii) of Theorem 19.2 is true for the following reasons. If the equality holds in (19.13), γ has the same minimizing properties as ξ, so that an R-parameterization of γ must be an extremal. As an extremal joining the endpoints of ξ in D_ξ^p, γ must be identical with ξ by Theorem 19.1.

Thus both (i) and (ii) of Theorem 19.2 are true.

<u>Geodesics on the Riemannian manifold</u> M_n. The theorems of this book are valid in the special case in which the W-preintegrands F, associated in §3 with the respective presentations $(\phi, U) \in \mathcal{D}M_n$, are defined by setting

$$F(u, r) = \mathfrak{f}(u, r) \tag{19.23}$$

where $\mathfrak{f}$ is the R-preintegrand associated with (ϕ, U) in §2. We shall term this the case (W = R). In this case we replace J by L of §2. In case (W = R) each "extremal" of L is an R-parameterized "geodesic" of the

R-manifold M_n. In the "local" coordinates of a presentation $(\phi, U) \in \mathcal{D}M_n$, the differential equations (7.1) of a geodesic on M_n take the form

$$
\left\{
\begin{aligned}
&\frac{d}{dt} \mathcal{L}_{r^i}(u, \dot{u}) = \mathcal{L}_{u^i}(u, \dot{u}) \\
&\frac{d}{dt} \mathcal{L}(u, \dot{u}) = 0 \ .
\end{aligned}
\right. \tag{19.24}
$$

The results of this section are valid with their special interpretation when (19.23) holds. In this case an "extremal" of L always means an R-parameterized geodesic on M_n. For this special case the following is true.

Theorem 19.3. In case (W = R) there exist open intervals of positive values κ, termed "field radii," for which the following is true.

Corresponding to points p and q in M_n such that $0 < d(p, q) \leq \kappa$ there exists a unique R-parameterized geodesic on M_n which joins p to q and affords a proper absolute minimum d(p, q) to R-length on M_n, relative to piecewise regular curves which join p to q on M_n.

Moreover the R-parameterized geodesics with R-lengths κ which issue from an arbitrary point p of M_n, bear no conjugate point of p and intersect in no point other than p.

In the case (W = R) the "preferred J-length" $\underset{\sim}{m}$ of Definition 19.4 can be taken as any field radius κ. Theorem 19.3 then follows from Theorems 19.1 and 19.2.

§20. J-distances $\Delta(p, q)$. §20 is a continuation of §19. The J-distance $\Delta(p, q)$ from a point $p \in M_n$ to a point $q \in M_n$ is here defined and the principal properties of $\Delta(p, q)$ derived. In particular basic relations between the J-distance $\Delta(p, q)$ and the R-distance $d(p, q)$ are derived.

The J-distance $\Delta(p, q)$ from p to q is taken as the G. L. B. (greatest lower bound) of J-lengths of admissible curves joining p to q. We are here assuming that each W-preintegrand F is nonsingular and positive definite. It follows from Theorem 19.2 that $\Delta(p, q)$ is bounded above. The J-distance $\Delta(p, q)$ may not equal $\Delta(q, p)$. See Bolza [1], p. 206, Beispiel XIV. The principal properties of $\Delta(p, q)$ will now be formulated.

Properties of $\Delta(p, q)$.

(a_1) If (p, q, r) is an ordered triple of points in M_n, then

$$\Delta(p, r) \leq \Delta(p, q) + \Delta(q, r) . \tag{20.1}$$

(a_2) If $\xi(p, q)$ is an elementary extremal with endpoints p and q, then

$$\Delta(p, q) = J(\xi(p, q)) \leq \underset{\sim}{m} \qquad (\underset{\sim}{m} \text{ of Def. 19.4}). \tag{20.2}$$

(a_3) If $\Delta(p, q) \leq \underset{\sim}{m}$ and $p \neq q$, a unique elementary extremal $\xi(p, q)$ exists with $J(\xi(p, q)) \leq \underset{\sim}{m}$.

(a_4) If $\Delta(p, q) \leq \underset{\sim}{m}$ and $p \neq q$ then $0 < d(p, q) \leq \underset{\sim}{\kappa}$ ($\underset{\sim}{\kappa}$ of Def. 19.4).

(a_5) *Corresponding to* $e > 0$ *there exists a constant* $\delta > 0$ *such that* $\Delta(p, q) < e$ if $0 < d(p, q) < \delta$.

(a_6) The mapping $(p, q) \to \Delta(p, q) : M_n \times M_n \to R$ is continuous.

(a_7) $\Delta(p, q) = 0$ if and only if $p = q$.

Proof of (a_1). The relation (20.1) follows at once from the definition of $\Delta(p, q)$.

Proof of (a_2). This property follows from Theorem 19.2 and the definition of $\Delta(p, q)$.

Proof of (a_3). Let γ be an elementary extremal of J issuing from p with J-length $\underset{\sim}{m}$ and let D^p_γ be the topological n-disc of elementary extremals issuing from p with J-lengths $\underset{\sim}{m}$. If $q \in D^p_\gamma$, (a_3) follows at once. If $q \notin D^p_\gamma$, any admissible curve h joining q to p must have a J-length exceeding $\underset{\sim}{m}$ by a positive constant independent of the choice of h. One infers that $\Delta(p, q) > \underset{\sim}{m}$, contrary to hypothesis. Thus (a_3) is true.

Proof of (a_4). By (a_3) the points (p, q) are joined by an elementary extremal $\xi(p, q)$ for which (20.2) holds. Let γ be an extremal extension of $\xi(p, q)$ with initial point p and J-length $\underset{\sim}{m}$. Let D^p_γ be the topological n-disc of (a_3). If $q \in D^p_\gamma$, (a_4) is true, since each elementary extremal issuing from $\underset{\sim}{0}$ in D^p_γ has an R-length $\leq \underset{\sim}{\kappa}$.

However, no other case can occur. Were $q \notin D^p_\gamma$ one infers as in the proof of (a_3) that $\Delta(p, q) > \underset{\sim}{m}$, contrary to hypothesis.

Proof of (a_5). To prove (a_5) two preliminary notational diversions are necessary.

Elementary geodesics. Consider the case in which the integral J is the R-integral L. Let σ be a field radius in this special case. Fix σ, and define an elementary geodesic on M_n as a geodesic of R-length at most σ.

In this special case Property (a_2) takes the following form. If $g(p, q)$ is an elementary geodesic joining p to q then

$$d(p,q) = L(g(p,q)) \; . \tag{20.3}$$

<u>A covering of</u> M_n. Returning to the case of a general J let

$$(\phi_i, U_i) \qquad (i = 1, \ldots, k) \tag{20.4}$$

be a set of presentations in $\mathcal{E}M_n$ with bounded domains whose ranges $\phi_i(U_i)$ cover M_n. Let βU_i be the boundary of U_i in R^n. For an arbitrarily small positive constant a let U_i^a be the open subset of U_i obtained by deleting from U_i all points of U_i at most a distance a from βU_i and set $\phi_i^a = \phi_i | U_i^a$. It is easy to prove the following. If $a_0 > 0$ is sufficiently small the ranges on M_n of the presentations

$$(\phi_i^{a_0}, U_i^{a_0}) \qquad (i = 1, \ldots, k)$$

cover M_n. More than that is true. Corresponding to any positive constant $\varepsilon < a_0$ there exists a positive constant $\eta < \sigma$ so small that each elementary geodesic $g(p, q)$ on M_n whose R-length is less than η has an antecedent in one of the coordinate domains U_i^ε. Let F_i and $\mathcal{f}_i$ be the W and R-preintegrands, respectively, associated with the presentation $(\phi_i^\varepsilon, U_i^\varepsilon)$. We come to the proof proper of (a_5).

Let a point pair (p, q) be prescribed on M_n with $0 < d(p, q) < \eta$. For some μ on the range $1, \ldots, k$, the elementary geodesic $g(p, q)$ has an antecedent under ϕ_μ^ε carried by U_μ^ε. By virtue of its definition

$$(20.5) \qquad \Delta(p,q) \leq J(g(p,q)) = \int_0^L F(u(s), \dot{u}(s))\, ds$$

where $s \to u(s)$ is an R-parameterization of $g(p,q)$ in U_μ^ε and $L = L(g(p,q))$. Note that $f(u(s), \dot{u}(s)) \equiv 1$. Set K_μ = maximum $F_\mu(u,r)$ subject to the condition that u be in $C\ell\, U_\mu^\varepsilon$ and $f_\mu(u,r) = 1$. Then (20.5) and (20.3) imply that

$$(20.6) \qquad \Delta(p,q) \leq K_\mu \int_0^L ds = K_\mu\, d(p,q) .$$

Since the constants $K_1, \ldots, K_k$ are finite in number, Property (a_5) follows.

Proof of (a_6). Let a constant $e > 0$ be prescribed. Let (p,q) and (p', q') be ordered pairs of points in M_n. We shall show that

$$(20.7) \qquad |\Delta(p,q) - \Delta(p',q')| < e$$

if for a sufficiently small $\delta > 0$

$$(20.8) \qquad d(p,p') < \delta, \quad d(q,q') < \delta.$$

By virtue of Property (a_1)

$$(20.9) \qquad \Delta(p,q) \leq \Delta(p,p') + \Delta(p',q') + \Delta(q',q) .$$

It follows from (a_5) that

(20.10) $$\Delta(p,p') < \frac{e}{2}, \quad \Delta(q',q) < \frac{e}{2}$$

if (20.8) holds with δ sufficiently small. In such a case (20.9) implies that $\Delta(p,q) - \Delta(p',q') < e$. One shows similarly that $\Delta(p',q') - \Delta(p,q) < e$ if (20.8) holds with δ sufficiently small. Hence (20.7) holds if (20.8) holds with δ sufficiently small.

Proof of (a_7). If $p = q$ then $\Delta(p,q) = 0$, as one proves by joining p to q by admissible curves of arbitrarily small J-length.

If $\Delta(p,q) = 0$, then $p = q$. Were $p \neq q$ let κ be a field radius less than $d(p,q)$ and let $b = m(\kappa)$ be the constant introduced in Def. 19.3. It follows from Theorem 19.2 that there exists a topological n-disc, say D_b^p of elementary extremals issuing from p with J-lengths b. The point $q \notin D_b^p$. It follows from Theorem 19.2 that any admissible curve which joins p to q on M_n has a J-length at least b. Hence $\Delta(p,q) \geq b > 0$ contrary to hypothesis. We infer that (a_7) is true.

Note. Property (a_5) can be supplemented by the following. Corresponding to $e > 0$ there exists a $\delta > 0$ such that $d(p,q) < e$ if $\Delta(p,q) < \delta$. This fact is not needed in what follows and hence will not be proved.

The J-lengths of elementary extremals. We have just seen that the mapping

(20.11) $$(p,q) \to \Delta(p,q) : M_n \times M_n \to R$$

is continuous. In general this mapping is not differentiable when $p = q$. We understand that $M_n \times M_n$ has the differential structure defined in Appendix III. The subspace Ω_n of $M_n \times M_n$ on which

$$(20.12) \qquad 0 < \Delta(p,q) < \underset{\sim}{m} \qquad (\underset{\sim}{m} \text{ of Def. 19.4})$$

is open. It is a differentiable submanifold of $M_n \times M_n$. We shall prove the following theorem.

__Theorem__ 20.1. _If_ Ω_n _is the subspace of_ $M_n \times M_n$ _on which_ (20.12) _holds, the mapping_

$$(20.13) \qquad (p,q) \to \Delta(p,q) : \Omega_n \to R$$

is of class C^∞.

We shall put Theorem 20.1 in an equivalent form. To that end first note the following. By virtue of statements (a_2), (a_3), and (a_7), condition (20.12) is equivalent to the condition that

$$(20.14) \qquad 0 < J(\xi(p,q)) < \underset{\sim}{m}$$

whenever $\xi(p,q)$ is an elementary extremal joining p to q on M_n. Since $\Delta(p,q) = J(\xi(p,q))$ under conditions (20.12) or (20.14), Theorem 20.1 is equivalent to the following.

__Theorem__ 20.2. _For elementary extremals_ $\xi(p,q)$ _for which_ (20.14) _holds, the mapping_

$$(p,q) \to J(\xi(p,q)) : \Omega_n \to R$$

is of class C^∞.

We shall prove Theorem 20.2 by verifying the form which it takes in local coordinates. To that end let $(\phi, U) \in \mathcal{L}M_n$ be given with an associated preintegrand F. Let

(20.15) $$\underset{\sim}{z} : s \to z(s) : [c', c''] \to U$$

be an extremal of J_F whose image under ϕ is an elementary extremal with J-length less than $\underset{\sim}{m}$. Theorem 20.2 is equivalent locally to the following lemma concerning $\underset{\sim}{z}$.

Lemma 20.1. *If* N' *and* N'' *are sufficiently small disjoint open neighborhoods in* U *of the respective endpoints* $z(c')$ *and* $z(c'')$ *of the extremal* (20.15), *the following is true.*

Arbitrary points $u' \in N'$ *and* $u'' \in N''$ *in* U *can be joined by unique elementary extremals* $\zeta(u', u'')$ *of* J_F *such that the mapping*

(20.16) $$(u', u'') \to J_F(\zeta(u', u'')) : N' \times N'' \to R$$

is of class C^∞.

That the mapping (20.16) is of class C^∞, if N' and N' are sufficiently small, will follow from Theorem* 5.1 of Morse [1].

---

*Theorem 5.1 of Morse [1] is true for a preintegrand f of class C^3. For present purposes it should be reformulated under the assumption that f is of class C^∞.

Without any loss of generality in proving Lemma 20.1 we can suppose that (ϕ, U) has been chosen so that the elementary extremal (20.15) reduces to the arc $\overrightarrow{c'c''}$ of the u^1-axis along which $s \equiv u^1$ and $c' \leq u^1 \leq c''$. An Euler preintegrand f can be introduced with values $f(x, y, p)$ defined by F as in (8.3). The "axial" extremal $\overrightarrow{c'c''}$ of J_F then has an x-parameterized "mate" g which is a conjugate point free extremal of J_f. Theorem 5.1 of Morse [1] can be formulated in terms of the mate g of $\underset{\sim}{z}$ and of extremals of J_f and will imply Lemma 20.1. Theorem 20.1 follows.

Canonical representations of elementary extremals. Theorem 20.2 concerns elementary extremals $\xi(p, q)$ for which the pair (p, q) satisfies (20.12). A point P of M_n on $\xi(p, q)$ will be canonically represented by a parameter t with a range $[0, 1]$ and with a value proportional to the J-length of $\xi(p, q)$ from p to P. Such a parameter equals 0 and 1 at p and q, respectively. The point P on $\xi(p, q)$ with canonical parameter t will be denoted by $P(t, p, q)$. We state a theorem.

Theorem 20.3. *If* Ω_n *is the subspace of pairs* $(p, q) \in M_n \times M_n$ *for which* (20.12) *holds, the canonical mapping*

$$(20.17) \qquad (t, p, q) \to P(t, p, q) : [0, 1] \times \Omega_n \to M_n$$

is of class C^∞.

The proof of this theorem is similar to the proof of Theorem 20.2 by means of Lemma 20.1. One makes similar local use of Theorem 5.1 of Morse [1].

Theorem 20.3 remains true if the condition (20.12) is replaced by the condition

(20.18) $$0 < \Delta(p, q) \leq \underset{\sim}{m} .$$

To verify this one has merely to redefine elementary extremals with $\underset{\sim}{m}$ replaced by a constant $\bar{\underset{\sim}{m}}$ slightly larger than $\underset{\sim}{m}$. Condition (20.12) then takes the form

$$0 < \Delta(p, q) < \bar{\underset{\sim}{m}} .$$

The modified theorem follows if $\bar{\underset{\sim}{m}} - \underset{\sim}{m}$ is sufficiently small.

§21. <u>Broken extremals joining A_1 to A_2</u>. Let distinct points A_1 and A_2 be prescribed on M_n. By an EB-extremal* joining A_1 to A_2 on M_n is meant a finite sequence of two or more elementary extremals of J successively joined to form a piecewise regular curve ζ joining A_1 to A_2 on M_n. We suppose that each EB-extremal joining A_1 to A_2 is parameterized, a priori, by R-length measured along ζ from A_1. An EB-extremal may be nonsimple. It may be defined by an extremal of J divided into a finite sequence of elementary extremals.

The search for extremals joining A_1 to A_2 can be reduced to a search among the EB-extremals which join A_1 to A_2 for those EB-extremals which have no "corners" at the points of junction of their successive elementary extremals. This will lead us to an extensive study of EB-extremals.

In Theorem 21.1 we show how this search reduces to the search for the critical points of suitably defined functions. In Theorem 21.2 we show that there exists an extremal joining A_1 to A_2 affording a minimum to J relative to admissible curves joining A_1 to A_2 in a prescribed <u>homotopy</u> class. A final theorem affirms that there exists an extremal joining A_1 to A_2 which affords a minimum to J relative to all admissible curves which join A_1 to A_2.

As in §20 we assume that each W-preintegrand is nonsingular and positive definite. After an introductory study of EB-extremals we shall prove the three principal theorems.

---

* EB-extremals abbreviates "elementary broken extremals."

Definition 21.0. EB-extremals $\zeta^{\nu}(\underset{\sim}{p})$ and their J-lengths $\mathcal{J}^{\nu}(\underset{\sim}{p})$. An EB-extremal γ joining A_1 to A_2 is defined by the sequence

$$A_1, p_1, p_2, \ldots, p_{\nu}, A_2 \qquad (\nu > 0) \tag{21.1}$$

of endpoints of the successive elementary extremals that are joined to define γ. We term the points $p_1, \ldots, p_{\nu}$, the vertices of γ and set $(p_1, \ldots, p_{\nu}) = \underset{\sim}{p}$. An EB-extremal with vertices $p_1, \ldots, p_{\nu}$ will be denoted by $\zeta^{\nu}(\underset{\sim}{p})$. The J-length of $\zeta^{\nu}(\underset{\sim}{p})$ will be denoted by $\mathcal{J}^{\nu}(\underset{\sim}{p})$. We shall refer to $\underset{\sim}{p}$ as a ν-tuple. It is a ν-tuple of vertices.

Definition 21.1. The vertex domain $Z^{(\nu)}$ of $\mathcal{J}^{\nu}$. Let $Z^{(\nu)}$ be the set of ν-tuples $\underset{\sim}{p}$ such that the sequence (21.1) defines an EB-extremal joining A_1 to A_2. We shall regard a ν-tuple $\underset{\sim}{p}$ as a point on the ν-fold product

$$M_n \times \cdots \times M_n = (M_n)^{\nu}$$

of M_n by itself and topologize $Z^{(\nu)}$ as a subspace of $(M_n)^{\nu}$. For products $(M_n)^{\nu}$ see Appendix III. The mapping

$$\underset{\sim}{p} \to \mathcal{J}^{\nu}(\underset{\sim}{p}) : Z^{(\nu)} \to R \tag{21.2}$$

will be called a vertex mapping into R of a vertex domain $Z^{(\nu)}$. Let $\overline{Z}^{(\nu)}$ denote the closure of $Z^{(\nu)}$ in $(M_n)^{\nu}$. The mapping (21.2) will presently be extended over $\overline{Z}^{(\nu)}$ so that $\overline{Z}^{(\nu)}$ will become the ultimate domain of $\mathcal{J}^{\nu}$.

The mapping $\mathcal{L}^{\nu}$. When the W-integral J is taken as the R-integral L, the mapping $\mathcal{J}^{\nu}$ will be denoted by $\mathcal{L}^{\nu}$.

Definition 21.2. The open subspace $\overset{\circ}{Z}^{(\nu)}$ of $Z^{(\nu)}$. The set $\overset{\circ}{Z}^{(\nu)}$ is the maximal open subset of $(M_n)^\nu$ included in $Z^{(\nu)}$. It is in fact the set of ν-tuples $\underset{\sim}{p} \in Z^{(\nu)}$ such that each elementary extremal in $\zeta^\nu(\underset{\sim}{p})$ has a J-length less than $\underset{\sim}{m}$. The set $Z^{(\nu)}$ may be empty a priori, if the integer ν is sufficiently small. However, neither $Z^{(\nu)}$ nor $\overset{\circ}{Z}^{(\nu)}$ is empty if $\nu > 0$ is so large that

$$(\nu+1)\underset{\sim}{m} > \Delta(A_1, A_2) \qquad (\underset{\sim}{m} \text{ of Def. 19.4}) \tag{21.3}$$

We shall suppose that the integer ν is so large that (21.3) holds. Under this assumption $\overset{\circ}{Z}^{(\nu)}$ is non-empty.

In Appendix III a differential structure of class C^∞ has been defined on the product manifold $(M_n)^\nu$ in terms of presentations in $\mathcal{L}M_n$. With this understood the following lemma has a well-defined meaning.

Lemma 21.1. The restriction of the vertex mapping $\mathcal{J}^\nu$ of (21.2) to the open subspace $\overset{\circ}{Z}^{(\nu)}$ of $(M_n)^\nu$ is of class C^∞.

Proof of Lemma 21.1. Let $\underset{\sim}{p} = (p_1, \ldots, p_\nu)$ be prescribed in $\overset{\circ}{Z}^{(\nu)}$. If one denotes A_1 and A_2 by p_0 and $p_{\nu+1}$, respectively, then

(21.4)* $$\mathcal{J}^\nu(\underset{\sim}{p}) = \sum_{i=1}^{\nu+1} J(\xi(p_{i-1}, p_i)) \, .$$

The i-th term on the right of (21.4) defines a mapping

$$\underset{\sim}{p} \to J(\xi(p_{i-1}, p_i)) \qquad (i = 1, \ldots, \nu+1)$$

---

* $\xi(p_{i-1}, p_i)$ is the elementary extremal on M_n with endpoints p_{i-1} and p_i.

of $\overset{c}{Z}^{(\nu)}$ into R. According to Theorem 20.2 each of these mappings is of class C^{∞} on $\overset{c}{Z}^{(\nu)}$. Since $\overset{c}{Z}^{(\nu)}$ is open, Lemma 21.1 is true.

By virtue of the following theorem, the search for extremals of J from A_1 to A_2 on M_n is reduced to the search for critical points of vertex functions $\mathcal{J}^{\nu}$ with vertex domains $\overset{c}{Z}^{(\nu)}$. Such functions exist for each integer ν satisfying (21.3). $\overset{c}{Z}^{(\nu)}$ is a differentiable manifold of dimension $n\nu$.

Theorem 21.1. *A necessary and sufficient condition that an EB-extremal $\zeta^{\nu}(\mathbf{q})$ with vertex set $\mathbf{q} \in \overset{c}{Z}^{(\nu)}$ be an extremal joining A_1 to A_2, is that the ν-tuple $\mathbf{q}$ be a critical point* of the vertex function $\mathcal{J}^{\nu}$ represented in terms of the νn-tuples of local coordinates of $\overset{c}{Z}^{(\nu)}$.*

Let $q \in M_n$ be an arbitrary one of the vertices $q_1, \ldots, q_{\nu}$ of the ν-tuple $\mathbf{q}$ and let q' and q'' be the vertices which precede and follow the vertex q, respectively, in the sequence $A_1, q_1, \ldots, q_{\nu}, A_2$. Then

$$\xi(q', q), \quad \xi(q, q'') \tag{21.5}$$

are elementary extremals which precede and follow the vertex q on the EB-extremal $\zeta^{\nu}(\mathbf{q})$. Let γ_q be the admissible curve on M_n obtained by tracing the elementary extremals (21.5) in the order written. To establish that the condition of Theorem 21.1 is sufficient, one has only to show that γ_q is an extremal of J, that is, that it has no corner at the point q at which the elementary extremals (21.5) are joined. To that end γ_q will be appropriately represented in local coordinates.

Among the presentations in $\mathcal{D}M_n$ of the type characterized in $(\underset{\sim}{A})$ of §19 there exists a presentation (ϕ, U) such that $\phi(\underset{\sim}{0}) = q$ and γ_q is the image

---

* The point $\mathbf{q}$ is a ν-tuple of vertices. It is represented locally by a νn-tuple of local coordinates of $\overset{c}{Z}^{(\nu)}$.

under ϕ of an admissible curve in U meeting the origin $\underset{\sim}{0}$ in U. Let N be an open neighborhood of the origin in U so small that the vertices q' and q'' do not meet $C\ell_{\phi}(N)$ on M_n and that

$$\Delta(q', \phi(u)) < \underset{\sim}{m}\ , \quad \Delta(\phi(u), q'') < \underset{\sim}{m} \qquad \text{(for } u \in N) . \tag{21.6}$$

Fix q' and q''. It follows from Theorem 20.1 that the mappings,

$$u \to J(\xi(q', \phi(u))), \quad u \to J(\xi(\phi(u), q'')) \qquad (u \in N) \tag{21.7}$$

of $N \to R$ are of class C^{∞}. Hence the mapping

$$u \to K(u) = J(\xi(q', \phi(u))) + J(\xi(\phi(u), q'')) \qquad (u \in N) \tag{21.8}$$

of N into R is of class C^{∞}. We continue with a proof of (α).

(α). *A necessary and sufficient condition that the R-parameterized join at q of the elementary extremals* (21.5) *be an extremal of J, is that the point* $u = \underset{\sim}{0}$ *be a critical point of the mapping* $u \to K(u)$ *of* (21.8).

Let J_F be the W-integral associated with the above presentation ϕ. $\xi(q', q)$ and $\xi(q, q'')$ will be represented in U by extremal arcs of J_F, say $\underset{\sim}{z}$ and $\hat{\underset{\sim}{z}}$. Let ρ and $\hat{\rho}$ be the n-tuples which are the "R-unitary directions" of the extremals $\underset{\sim}{z}$ and $\hat{\underset{\sim}{z}}$, respectively, at $u = \underset{\sim}{0}$. With u restricted to N, we shall prove the following:

$$\frac{\partial}{\partial u^i} J(\xi(q', \phi(u)))\Big|^{u=\underset{\sim}{0}} = F_{r^i}(\underset{\sim}{0}, \rho) \qquad (i = 1, \ldots, n) \tag{21.9}$$

(21.10) $$\frac{\partial}{\partial u^i} J(\xi(\phi(u), q''))\Big|^{u=\underset{\sim}{0}} = -F_{r^i}(\underset{\sim}{0}, \hat{\rho}) \qquad (i = 1, \ldots, n)$$

Proof of (21.9). For $u \in N$ and N sufficiently small, the initial point u_0 of $\underset{\sim}{z}$ and an arbitrary point $u \in N$ can be joined in U by an elementary extremal, say $\underset{\sim}{z}^u$ of J_F. Recall that the extremal arc, $\underset{\sim}{z}^u$ is parameterized by R-length s. Let λ^u be the extremaloid (Def. 7.1) obtained from $\underset{\sim}{z}^u$ by giving $\underset{\sim}{z}^u$ an RL-parameter t with range $[0,1]$. Recall that

(21.11) $$J_F(\underset{\sim}{z}^u) = J_F(\lambda^u) = J(\xi(q', \phi(u))) \qquad (u \in N).$$

For N sufficiently small and $u \in N$, λ^u is given by a mapping

(21.12) $$t \to X(t, u) : [0,1] \to U$$

such that $X(0,u) \equiv u_0$, $X(1,u) \equiv u$, and the mapping

$$(t, u) \to X(t, u) : [0,1] \times N \to U$$

is of class C^∞. for $u \in N$

$$\frac{\partial}{\partial u^i} J_F(\lambda^u) = \frac{\partial}{\partial u^i}\int_0^1 F(X(t,u), X_t(t,u))\, dt,$$

from which it follows that for $(i = 1, \ldots, n)$

$$\frac{\partial}{\partial u^i} J_F(\lambda^u)\Big|^{u=\underset{\sim}{0}} = F_{r^j}(\underset{\sim}{0}, \rho) X^j_{u^i}(1, u)\Big|^{u=\underset{\sim}{0}} = F_{r^i}(\underset{\sim}{0}, \rho).$$

Thus (21.9) holds. The proof of (21.10) is similar.

Completion of proof of (a). The condition of (a) is sufficient. If $u = \underset{\sim}{0}$ is a critical point of the mapping (21.8), then

$$F_{r^i}(\underset{\sim}{0}, \rho) - F_{r^i}(\underset{\sim}{0}, \hat{\rho}) = 0 \qquad (i = 1, \ldots, n) \tag{21.13}$$

by virtue of (21.9) and (21.10). It follows that

$$0 = \rho^i\left(F_{r^i}(\underset{\sim}{0}, \rho) - F_{r^i}(\underset{\sim}{0}, \hat{\rho})\right) = F(\underset{\sim}{0}, \rho) - \rho^i F_{r^i}(\underset{\sim}{0}, \hat{\rho}) = \mathcal{E}_F(\underset{\sim}{0}, \hat{\rho}, \rho) \tag{21.14}$$

in accord with Definition 5.16 of $\mathcal{E}_F$. Since ρ and $\hat{\rho}$ are R-unitary n-tuples at $u = \underset{\sim}{0}$, Lemma 18.1 implies that $\rho = \hat{\rho}$. This is under our hypothesis that each preintegrand be nonsingular and positive regular. Hence the R-parameterized join of the elementary extremals (21.5) is an extremal.

One proves that the condition of (a) is necessary as follows. If the R-parameterized join of the extremal arcs (21.5) is an extremal, then $\rho = \hat{\rho}$ and (21.13) holds. It follows from (21.9) and (21.10) that $u = \underset{\sim}{0}$ is a critical point of the mapping (21.8).

Thus statement (a) is true. Theorem 21.1 follows.

Before coming to the homotopy Theorem 21.2, it is technically necessary to further study broken extremals.

Singular broken extremals. It is necessary to examine the closure $\bar{Z}^{(\nu)}$ of $Z^{(\nu)}$ and the nature of the components* of $Z^{(\nu)}$. If, for example, M_n is a torus, $Z^{(\nu)}$ will have a finite number of pathwise components, each a subspace

*We refer to maximal pathwise connected subsets of $Z^{(\nu)}$ as components of $Z^{(\nu)}$.

of $(M_n)^\nu$. As $\nu \uparrow \infty$ the number of these components will increase without limit. The points A_1 and A_2 are distinct and fixed. The number of components of $Z^{(\nu)}$ depends on the choice of A_1 and A_2.

$Z^{(\nu)}$ may not be closed in $(M_n)^\nu$. If $\underset{\sim}{q} = (q_1, \ldots, q_\nu)$ is in $\bar{Z}^{(\nu)}$ but not in $Z^{(\nu)}$ at least two successive points of M_n in the sequence

$$A_1, q_1, \ldots, q_\nu, A_2 \tag{21.15}$$

must be identical. Such a ν-tuple $\underset{\sim}{q}$ will be called a <u>singular</u> ν-tuple of $\bar{Z}^{(\nu)}$.

<u>Definition</u> 21.3. <u>Singular broken extremals</u> $\zeta^\nu(\underset{\sim}{q})$. If $\underset{\sim}{q}$ is a singular ν-tuple of $\bar{Z}^{(\nu)}$, successive points in the sequence (21.15) which are not identical are the endpoints of an elementary extremal. The join of these elementary extremals will still be denoted by $\zeta^\nu(\underset{\sim}{q})$. As a representation of this join, $\zeta^\nu(\underset{\sim}{q})$ will be termed a <u>singular broken extremal</u>. A singular broken extremal $\zeta^\nu(\underset{\sim}{q})$ has fewer than ν distinct vertices. By hypothesis $\nu > 1$. A singular broken extremal $\zeta^\nu(\underset{\sim}{q})$ can even reduce to an elementary extremal $\xi(A_1, A_2)$. However, this can happen only if $\Delta(A_1, A_2) \leq \underset{\sim}{m}$. The value of J along $\zeta^\nu(\underset{\sim}{q})$ is called its J-<u>length</u>.

<u>Definition</u> 21.4. An extension of $\mathcal{J}^\nu$ over $\bar{Z}^{(\nu)}$. For $\underset{\sim}{p} \in Z^{(\nu)}$, $J^\nu(\underset{\sim}{p})$ is well-defined. If $\underset{\sim}{q}$ is a singular ν-tuple of $\bar{Z}^{(\nu)}$ one sets

$$\mathcal{J}^\nu(\underset{\sim}{q}) = \lim_{\underset{\sim}{p} \to \underset{\sim}{q}} \mathcal{J}^\nu(\underset{\sim}{p}) \qquad (\underset{\sim}{p} \in Z^{(\nu)}). \tag{21.16}$$

The continuity of $\Delta(p, q)$, established in §20, and Property (a_2) of $\Delta(p, q)$ imply that the limit (21.16) is independent of the approach of $\underset{\sim}{p}$ to $\underset{\sim}{q}$ from $Z^{(\nu)}$. The continuity of $\mathcal{J}^{\nu}$ on $\overline{Z}^{(\nu)}$ is also implied. The relation

$$(21.17) \qquad \mathcal{J}^{\nu}(\underset{\sim}{q}) = J(\zeta^{\nu}(\underset{\sim}{q}))$$

holds even when $\underset{\sim}{q}$ is singular. Whether singular or nonsingular, $\zeta^{\nu}(\underset{\sim}{q})$ shall be parameterized by R-length s.

Pathwise components of $Z^{(\nu)}$. Let $\zeta(Z^{(\nu)})$ denote the set of EB-extremals $\zeta^{\nu}(\underset{\sim}{p})$, as $\underset{\sim}{p}$ ranges over $Z^{(\nu)}$. These EB-extremals join A_1 to A_2. If $\underset{\sim}{p}'$ and $\underset{\sim}{p}''$ are distinct ν-tuples in $Z^{(\nu)}$, then the EB-extremals $\zeta^{\nu}(\underset{\sim}{p}')$ and $\zeta^{\nu}(\underset{\sim}{p}'')$ are homotopic relative to the set $\zeta(Z^{(\nu)})$, of EB-extremals, if and only if $\underset{\sim}{p}'$ and $\underset{\sim}{p}''$ are ν-tuples in the same pathwise component X of $Z^{(\nu)}$. See Definitions 7.5 and 7.6 for relative homotopy.

Let X be a pathwise component of $Z^{(\nu)}$. Note that $\overline{X}$ is a pathwise connected subset of $(M_n)^{\nu}$ and that the corresponding set $\zeta(\overline{X})$ of broken extremals joining A_1 to A_2 is self-homotopic in the sense of Definition 7.6. We shall circumvent the difficulties arising from the existence of singular ν-tuples in $\overline{Z}^{(\nu)}$ by the use of a special kind of ν-tuple in $Z^{(\nu)}$ termed J-normal.

Definition 21.5. J-normal ν-tuples. A ν-tuple $\underset{\sim}{q} \in Z^{(\nu)}$ is termed J-normal if the elementary extremals which are joined to define $\zeta^{\nu}(\underset{\sim}{q})$ are equal in J-length.

Concerning the existence of J-normal ν-tuples we shall prove the following lemma.

<u>Lemma</u> 21.2. *Let X be a component of* $\overline{Z}^{(\nu)}$. *Corresponding to each* ν*-tuple* $\underset{\sim}{p}^0 \in X$, *there exists a J-normal* ν*-tuple* $\underset{\sim}{q} \in X$ *such that* $\mathcal{J}^\nu(\underset{\sim}{q}) \leq \mathcal{J}^\nu(\underset{\sim}{p}^0)$.

Lemma 21.2 is a consequence of the following.

<u>Lemma</u> 21.3. *Let X be a pathwise component of* $\overline{Z}^{(\nu)}$ *and set*

$$\text{(21.18)} \qquad \underset{\underset{\sim}{p} \in X}{\text{infimum}} \; \mathcal{J}^\nu(\underset{\sim}{p}) = c \; .$$

There then exists a J-normal ν*-tuple* $\underset{\sim}{q} \in X$ *which defines an extremal* $\gamma = \zeta^\nu(\underset{\sim}{q})$ *joining* A_1 *to* A_2 *such that* $J(\gamma) = c$.

There exists an infinite sequence $(\underset{\sim}{p}^{(m)})$, $m = 1, 2, \ldots,$ of ν-tuples $\underset{\sim}{p}^{(m)}$ in X and a ν-tuple $\hat{\underset{\sim}{q}} = (\hat{q}_1, \ldots, \hat{q}_\nu) \in X$ such that

$$\text{(21.19)} \qquad \lim_{m\uparrow\infty} \mathcal{J}^\nu(\underset{\sim}{p}^{(m)}) = c \; ; \; \lim_{m\uparrow\infty} p^{(m)} = \hat{\underset{\sim}{q}} \; .$$

The broken extremal $\zeta^\nu(\hat{\underset{\sim}{q}})$, if properly parameterized, will be an extremal γ which joins A_1 to A_2. Let $\underset{\sim}{q} = (q_1, \ldots, q_\nu)$ be a J-normal ν = tuple which divides γ into $\nu + 1$ successive subcurves of equal J-length. It remains to prove the following:

(i) *The J-normal* ν*-tuple* $\underset{\sim}{q}$ *is in* X.

To establish (i) let any point on γ with parameter s_0 be assigned a J-coordinate equal to the J-length of the subcurve of γ on which $0 \leq s \leq s_0$. As the time t increases from 0 to 1 let the ν-tuple $\hat{\underset{\sim}{q}}$ be replaced by a ν-tuple $\underset{\sim}{q}^t = (q_1^t, \ldots, q_\nu^t)$ whose vertices q_i^t move along γ from $\hat{\underset{\sim}{q}}_i$ to $\underset{\sim}{q}_i$, respectively. If this movement is such that the J-coordinate of q_i^t changes at a constant rate, successive points in the sequence

$$A_1, q_1^t, q_2^t, \ldots, q_\nu^t, A_2 \qquad (0 < t \leq 1) \tag{21.20}$$

will be distinct and define elementary extremals. Statement (i) follows.

Thus Lemma 21.3 is true.

The homotopy theorem. We come to a theorem on the existence of an extremal in a prescribed homotopy class of admissible curves joining A_1 to A_2. Let $((A_1, A_2))$ denote the set of all admissible curves joining A_1 to A_2.

Theorem 21.2. *Let* γ *be prescribed in* $((A_1, A_2))$ *and let* $h(\gamma)$ *be the class of curves in* $((A_1, A_2))$ *homotopic to* γ *relative to* $((A_1, A_2))$. *Set*

$$c = \underset{z \in H(\gamma)}{\text{infimum}}\ J(z)\ . \tag{21.21}$$

There then exists an extremal g *in the homotopy class* $h(\gamma)$ *such that* $J(g) = c$.

From (21.21) we infer the existence of an infinite sequence $(z^{(m)})$ of curves in the homotopy class $h(\gamma)$ such that

$$\lim_{m \uparrow \infty} J(z^{(m)}) = c\ . \tag{21.22}$$

We can suppose that for some $b > c$, $J(z^{(m)}) < b$ for all m. There then exists an integer ν so large that each curve $z^{(m)}$ can be subdivided into ν consecutive subcurves of equal J-length less than $\underset{\sim}{m}$. Let

$$A_1, p_1^{(m)}, \ldots, p_\nu^{(m)}, A_2 \tag{21.23}$$

be the sequence of endpoints in M_n *of these subcurves of* $z^{(m)}$. *It follows* from Theorem 19.2 that

$$\mathcal{J}^\nu(\underset{\sim}{p}^{(m)}) \leq J(z^{(m)}) \qquad (m = 1, 2, \ldots). \tag{21.24}$$

Let X_m be a pathwise component of $\overline{Z}^{(\nu)}$ which contains the ν-tuple $\underset{\sim}{p}^{(m)}$. According to Lemma 21.2 there exists a J-normal ν-tuple $\underset{\sim}{q}^{(m)}$ in X_m such that $\mathcal{J}^\nu(q^{(m)}) \leq \mathcal{J}^\nu(\underset{\sim}{p}^{(m)})$. Each curve $\zeta^\nu(\underset{\sim}{q}^{(m)})$ is in the given homotopy class $h(\gamma)$.

A subsequence of the sequence $(\underset{\sim}{q}^{(m)})$ will converge to a J-normal ν-tuple $\underset{\sim}{q}$. It is clear that $\mathcal{J}^\nu(\underset{\sim}{q}) = c$ and that $\underset{\sim}{q}$ is accordingly a critical point of $\mathcal{J}^\nu$. Hence by Theorem 21.1 $\zeta^\nu(\underset{\sim}{q})$ is an extremal $\hat{\gamma}$ of J. One sees that $\hat{\gamma}$ is A_1A_2-homotopic to γ.

Theorem 21.2 follows.

The methods of proof of Lemma 21.3 and Theorem 21.2 lead to a proof of the following theorem.

<u>Theorem</u> 21.3. <u>Let</u> $((A_1, A_2))$ <u>denote the set of piecewise regular curves joining</u> A_1 <u>to</u> A_2. <u>Set</u>

$$\underset{z \in ((A_1, A_2))}{\text{infimum}} \quad J(z) = c .$$

<u>There then exists an extremal</u> γ <u>of</u> J <u>joining</u> A_1 <u>to</u> A_2, <u>such that</u> $J(\gamma)$ <u>gives a minimum</u> c <u>to</u> J <u>relative to all admissible curves joining</u> A_1 <u>to</u> A_2.

Chapter 8

Nonsimple extremals

§22. Tubular mappings into M_n. Prior to Part IV, the principal theorems have been restricted to extremal arcs, that is extremals that are simple. Extremals γ of J on M_n which are self-intersecting cannot be excluded in a global theory. One must define conjugate points on such an extremal γ and give a precise definition of the classes of curves relative to which a properly conditioned extremal γ gives a proper minimum to J. In such a study a most relevant property of an extremal of J on M_n is that it is regular. We, accordingly, begin with a regular curve*,

$$s \to \zeta(s) : [a, b] \to M_n \tag{22.1}$$

on M_n, parameterized by R-length s, measured from a convenient point of ζ and understand that ζ may be nonsimple as well as simple.

The curve ζ may be supposed given by representations of successive subarcs in a finite sequence of overlapping coordinate domains of presentations in $\mathcal{D}M_n$. A first objective is to show that, if ζ is both simple and regular, it is representable in the coordinate domain U of a single presentation $(\phi, U) \in \mathcal{D}M_n$. A theorem of this type is known to differential geometers in various forms. A broader objective is to appropriately represent neighborhoods of nonsimple regular curves.

To this end we shall define a special class of differentiable mappings

---

* All regular curves in this section are supposed to be of class C^∞. Extremals of J have this property.

into M_n, termed tubular mappings. Tubular mappings may or may not be in $\mathcal{D}M_n$.

Notation. Let $u^1, \ldots, u^n$ be rectangular coordinates in R^n. Let $[\alpha, \beta]$ be a finite interval on the u^1-axis, $\overrightarrow{\alpha, \beta}$ the corresponding arc, parameterized by u^1. Let $|\alpha, \beta|$ denote the carrier of $\overrightarrow{\alpha, \beta}$ in R^n.

Definition 22.1. Tubular domains. An open relatively compact neighborhood of $|\alpha, \beta|$ in R^n will be termed a tubular domain with central arc $\overrightarrow{\alpha, \beta}$. It will be denoted by $U(\alpha, \beta)$ or $V(\alpha, \beta)$.

Definition 22.2 (i). Tubular mappings. A mapping

$$u \to \pi(u) : U(a, b) \to M_n \tag{22.2}$$

of class C^∞ will be called a tubular mapping.

The tubular mappings, actually to be used, belong to one of the two following special classes.

Definition 22.2 (ii). Simple tubular mappings. These are tubular mappings (22.2) which define a presentation $(\pi, U(a, b)) \in \mathcal{D}M_n$.

Definition 22.2 (iii). Regular tubular mappings. These are mappings (22.2) such that $[a, b]$ can be decomposed into a finite sequence

$$[c_1, c_2], [c_2, c_3], \ldots, [c_\mu, c_{\mu+1}] \qquad (\mu > 1) \tag{22.3}$$

of intervals with the respective overlapping tubular neighborhoods,

$$U(c_1, c_2), U(c_2, c_3), \ldots, U(c_\mu, c_{\mu+1}) , \tag{22.4}$$

of which $U(a,b)$ is the union and on the ith of which the restriction

$$\pi | U(c_i, c_{i+1}) \qquad (i = 1, \ldots, \mu) \tag{22.5}$$

is a <u>simple</u> tubular mapping.

The restrictions (22.5) are in $\mathcal{D}M_n$ and will be termed <u>subordinate</u> to π.

Simple tubular mappings are regular tubular mappings, but the converse is not true. Regular tubular mappings are locally diffs. For a regular tubular mapping of form (22.2), the curve* $\pi \cdot \overrightarrow{a,b}$ on M_n is regular and is called the <u>central image</u> of π in M_n. Corresponding to the two fundamental classes of tubular mappings, there are two fundamental theorems.

<u>Theorem</u> 22.1. <u>Corresponding to a simple regular arc*</u> ζ <u>on</u> M_n <u>on which</u> $a \leq s \leq b$, <u>there exists a simple tubular mapping</u>,

$$u \to \pi(u) : U(a,b) \to M_n \tag{22.6}$$

<u>such that</u> $\zeta(c) = \pi(c, 0, \ldots, 0)$ <u>for each point</u> $(c, 0, \ldots, 0) \in |a,b|$.

<u>Theorem</u> 22.2. <u>Corresponding to a regular curve*</u> ζ <u>on</u> M_n <u>on which</u> $a \leq s \leq b$, <u>there exists a regular tubular mapping</u>,

$$u \to \pi(u) : U(a,b) \to M_n , \tag{22.7}$$

<u>such that</u> $\zeta(c) = \pi(c, 0, \ldots, 0)$ <u>for each point</u> $(c, 0, \ldots, 0) \in |a,b|$.

Theorem 22.2 will be proved in Appendix I. It has a simple corollary.

---

* ζ is of class C^∞ by hypothesis, - an extremal of J, if one pleases.

Corollary 22.1. *If* $[\alpha, \beta]$ *is any subinterval of the interval* $[a, b]$ *of Theorem* 22.2 *such that* ζ *is simple for* $\alpha \leq s \leq \beta$, *then, if* $U(\alpha, \beta)$ *is a sufficiently small open tubular neighborhood of* $|\alpha, \beta|$ *in* R^n, *the restriction* $\pi | U(\alpha, \beta)$ *is a simple tubular mapping such that* $\zeta(c) = \pi(c, 0, \ldots, 0)$ *for each point* $(c, 0, \ldots, 0) \in |\alpha, \beta|$

This corollary is a consequence of two facts. A regular tubular mapping is locally a diff, while $\zeta | [\alpha, \beta]$, by hypothesis, is both simple and regular. If, then, $U(\alpha, \beta) \subset U(a, b)$ and if $U(\alpha, \beta)$ is a sufficiently small open neighborhood of $|\alpha, \beta|$ in R^n, $\pi | U(\alpha, \beta)$ must be a biunique mapping into M_n. Such a mapping is in $\mathcal{D}M_n$, since $U(\alpha, \beta) \subset U(a, b)$ and π is compatible with the presentations in $\mathcal{D}M_n$ "subordinate" to π.

We record the following consequence of Corollary 22.1.

Theorem 22.1 *is a corollary of Theorem* 22.2.

Definition 22.3. *Extendable diffs.* If Θ is a diff of a relatively compact open subset V of R^n into M_n or R^n, a restriction $\Theta | U$ of Θ to an open subset U of V with closure $\bar{U} \subset V$, will be termed *extendable*.

Extendable diffs have many uniform properties of differentiability. If, for example, H is an extendable diff of U into R^n of the form $u \to H(u)$, the Jacobian of H will be bounded from zero on U. Extendable diffs will be used in Appendix I in the proof of Theorem 22.2.

Definition 22.4. *Axial representations.* When Theorem 22.2 holds, we say that the regular curve ζ on M_n (possibly an extremal of J) is *axially represented* under π by the arc $\overrightarrow{a, b}$ of the u^1-axis.

§23. W-preintegrands F^{π}, π a regular tubular mapping. With each presentation $(\phi, U) \in \mathcal{D}M_n$ we have associated a unique W-preintegrand F. With each regular tubular mapping π we shall here associate a unique pre-integrand to be denoted by F^{π}.

Definition 23.1. The W-preintegrand F^{π}. If the mapping π has the tubular domain U(a, b), the domain of F^{π} shall be the set of 2n-tuples

$$(u, r) \in U(a, b) \times \dot{R}_n \tag{23.1}$$

For each domain $U(\alpha, \beta)$ of a presentation (22.5) in $\mathcal{D}M_n$ which is "subordinate" to π (Def. 22.2(iii)), and which has a W-preintegrand F, we shall set

$$F^{\pi}(u, r) = F(u, r) \qquad [(u, r) \in U(\alpha, \beta) \times \dot{R}^n] \tag{23.2}$$

F^{π} is thereby overdefined, but consistently, by virtue of the compatibility conditions of §3 on W-preintegrands F_i associated with the respective presentations (22.5) "subordinate" to π. We note that the W-preintegrands F^{π} are of class C^{∞} on their domains.

R-preintegrands f^{π}. We have associated a W-preintegrand F^{π} with each regular tubular mapping π. We similarly associate an R-preintegrand f^{π} with each regular tubular mapping π. For this purpose use is made of the R-preintegrands f_i associated with the respective presentations (22.5) subordinate to π.

The W-integral $J_{F^{\pi}}$. Let a regular tubular mapping π be given as in Definition 22.2 (iii). Corresponding to each piecewise regular curve

(23.3) $$\underset{\sim}{u} : t \to u(t) : [a', a''] \to U(a, b)$$

one sets

(23.4) $$J_{F^\pi}(u) = \int_{a'}^{a''} F^\pi(u(t), \dot{u}(t))\, dt ,$$

thereby defining the W-integral J_{F^π}.

Recall the definition of the Euler-Riemann differential equations (7.0). The Euler-Riemann differential equations of J_{F^π} are similarly defined as conditions on regular curves $\underset{\sim}{u}$ in U(a, b). These equations have the form

(23.5)' $$\frac{d}{dt} F^\pi_{r^i}(u, \dot{u}) = F^\pi_{u^i}(u, \dot{u}) \qquad (i = 1, \ldots, n)$$

(23.5)'' $$\frac{d}{dt} \mathcal{F}^\pi(u, \dot{u}) = 0$$

Let $U(\alpha, \beta)$ be the tubular domain of any one of the presentations (22.5) in $\mathcal{E}M_n$ "subordinate" to π and F the associated W-preintegrand. When $\underset{\sim}{u}$ is restricted to $U(\alpha, \beta)$, the conditions (22.5) reduce to the Euler-Riemann equations of J_F.

Axial representations of extremals of J. Each extremal γ of J on M_n of finite R-length ρ can be axially represented in the sense of Definition 22.4. Theorem 22.2 implies that there exists a regular tubular mapping

(23.6) $$u \to \pi(u) : U(o, \rho) \to M_n$$

such that $\pi \cdot \overrightarrow{o, \rho} = \gamma$. This theorem has the following corollary.

Corollary 23.1 of Theorem 22.2. *If a regular curve γ on M_n of R-length ρ is the central image* $\pi \cdot \overrightarrow{o,\rho}$ of a regular tubular mapping π of form (23.6), the arc $\overrightarrow{o,\rho}$ is an extremal of J_{F^π}, if and only if γ is an extremal of J.*

If π is a simple tubular mapping and hence in $\mathcal{E}M_n$, the corollary is a trivial consequence of Definition 7.2 of an extremal arc of J on M_n. If π is a general regular tubular mapping, the characterization of the Euler-Riemann equations of J_{F^π}, immediately following (23.5), shows that any sufficiently short subarc h of $\overrightarrow{o,\rho}$ is an extremal of J_{F^π} if and only if $\pi \cdot h$ is an extremal of J, the integral on M_n.

The corollary follows.

Conjugate points on an extremal γ of J. Conjugate points have not yet been defined on γ when γ is nonsimple. We give a definition of conjugate points on γ which is equivalent to the definition in §10 when γ is simple.

To that end we suppose that γ has the R-length ρ and is the central image, $\pi \cdot \overrightarrow{o, \rho}$ of a regular tubular mapping π : (23.6). By the above corollary, the subarc $\overrightarrow{o, \rho}$ of the u^1-axis is an extremal of J_{F^π} with $u^1 = s$ thereon. To define conjugate points on γ we first define conjugate points on $\overrightarrow{o, \rho}$ as an extremal of J_{F^π}.

---

* For the meaning of $\pi \cdot \overrightarrow{o, \rho}$ see footnote to Lemma 3.1. Recall that u_1 is the parameter of the arc $\overrightarrow{o, \rho}$.

Conjugate points on $\overrightarrow{o,\rho}$. A proper polar family $\underset{\sim}{\Phi}$ of extremal arcs of J_{F^π} on $U(o,\rho)$, is introduced as in §9, with the arc $\overrightarrow{o,\rho}$ as "central" extremal. The tubular domain $U(o,\rho)$ plays the role of U in §9. As in §9 one has a C^∞-mapping,

$$(s,a) \to \Phi(s,a) : [o,\rho] \times B_m \to U(o,\rho) \qquad (23.7) \quad (\text{cf. } (9.2)')$$

whose partial mappings, for a fixed in B_m, are extremal arcs of J_{F^π} in $U(o,\rho)$. Conjugate points of a point $u^1 = s = c$ on $\overrightarrow{o,\rho}$ are defined, as in §10, by the zeros of $D^c_\Phi(s)$, each with its multiplicity.

We return to γ. If $u^1 = c^*$ is conjugate to $u^1 = c$ on $\overrightarrow{o,\rho}$ with multiplicity μ, we understand that, on γ, the point $s = c^*$ is conjugate on γ to the point $s = c$ with the above multiplicity μ Cf. Corollary 10.2.

A supplementary lemma. If an extremal γ of J of R-length ρ is given on M_n, regular tubular mappings π with γ as the central extremal $\pi \cdot \overrightarrow{o,\rho}$ of π are not unique. Our definition of conjugate points on γ accordingly requires the following supplement.

Lemma 23.1. *Conjugate points on an extremal γ of J on M_n of R-length ρ, defined as above, when γ is the central image $\pi \cdot \overrightarrow{o,\rho}$ of a regular tubular mapping π, are identical with the conjugate points on γ, similarly defined when γ is the central image $\hat{\pi} \cdot \overrightarrow{o,\rho}$ of a second regular tubular mapping $\hat{\pi}$.*

Let $u^1, \ldots, u^n$ be rectangular coordinates of points $u \in R^n$ and $v^1, \ldots, v^n$ be similar rectangular coordinates of points $v \in R^n$. We understand that u and v represent the same point in R^n if and only if $u = v$. Let the tubular domain of π be given as a set of points u denoted by $U(o, \rho)$. Let the tubular domain of $\hat{\pi}$ be given as a set of points v denoted by $V(o, \rho)$. Lemma 23.1 is a consequence of the following statement.

(i) If N is a sufficiently small open neighborhood of $|o, \rho|$ in $U(o, \rho)$, there exists a unique diff

$$u \to \Theta(u) : N \to V(o, \rho) , \tag{23.8}$$

of class C^∞, of N onto an open neighborhood $\Theta(N)$ of $|o, \rho|$ in $V(o, \rho)$, such that Θ leaves the points of $|o, \rho|$ fixed and implies that $\hat{\pi}(\Theta(u)) \equiv \pi(u)$ for $u \in N$.

To verify (i), recall that both of the mappings π and $\hat{\pi}$ are diffs locally. If one takes into account the fact that γ is regular, so that any sufficiently short arc of γ is simple, one sees that, if N is sufficiently small, (i) is true.

Granting (i), one verifies Lemma 23.1 as follows. Any proper polar family $\underset{\sim}{\Phi}$ of extremals of J_{F^π}, issuing from a point $u^1 = c$ on $|o, \rho|$ in $U(o, \rho)$ with initial directions sufficiently near the direction of $\overrightarrow{o, \rho}$, will be mapped by Θ onto a proper polar family $\Theta\underset{\sim}{\Phi}$ of extremals of $J_{F^{\hat{\pi}}}$ in $V(o, \rho)$ issuing from the point $v^1 = c$ on the v^1-axis. One has $\pi(\Phi) = \hat{\pi}(\Theta(\Phi))$ on M_n, so that $\underset{\sim}{\Phi}$ and $\Theta(\underset{\sim}{\Phi})$ define the same extremals of J on M_n.

Lemma 23.1 follows. (Cf. proof of Lemma 9.4.)

The following theorem is a first example of how one obtains theorems on extremals γ of J on M_n that may be self-intersecting, from theorems on simple extremals. The theorem on simple extremals to be used is Theorem 18.1.

Theorem 23.1. Let γ be an extremal of J on M_n of R-length ρ and let

$$u \to \pi(u) : U(o,\rho) \to M_n \tag{23.9}$$

be a regular tubular mapping π such that $\gamma = \pi \cdot \overrightarrow{o,\rho}$. If then J is positive regular in the sense of Definition 14.3, if γ is free of conjugate points and if N is any sufficiently small open neighborhood of $|o,\rho|$ in $U(o,\rho)$, the following is true.

The extremal γ affords a proper minimum to J relative to the images under π of admissible curves in N which join the endpoints of the arc $\overrightarrow{o,\rho}$.

By hypothesis* each W-preintegrand F is positive regular. It follows that the special W-preintegrand F^π is positive regular. The arc $\overrightarrow{o,\rho}$ is an extremal of J_{F^π} by Corollary 23.1.

Although F^π is not a W-preintegrand associated with a presentation $(\phi, U) \in \mathcal{D}M_n$, nevertheless Theorem 18.1 and its proof are valid when F is replaced by F^π and $\underset{\sim}{z}$ by $\overrightarrow{o,\rho}$. Hence if N is a sufficiently small open neighborhood of $|o,\rho|$, the extremal $\overrightarrow{o,\rho}$ affords a proper minimum to J_{F^π}

---

* For simpler hypotheses see Theorem 15.1.

relative to admissible curves h which join the endpoints of $\overrightarrow{o, \rho}$ in N.

Theorem 23.1 now follows from the relation

$$J(\pi \cdot h) = J_{F^{\pi}}(h) . \tag{23.10}$$

The relation (23.10) is clearly true when π is in $\mathcal{D}M_n$ and follows in the general case, on making use of the Definition (23.2) of F^{π}.

Separation Theorem on nonsimple extremals. The theorems on conjugate points on extremal arcs of J in §10 yield similar theorems on extremals γ that are no longer simple. It must be understood that conjugate points on γ are defined in terms of R-length s along γ and not in terms of points on $|\gamma|$ which may be represented by several different values of s. A theorem of particular importance in the global theory is an extension of Theorem 10.5 stated as follows.

Theorem 23.2. *Let γ be an extremal of J on M_n (simple or non-simple) on which $0 \leq s \leq s_1$. If $c' < c''$ are two values of s in $[o, s_1]$, the count of conjugate points of $s = c'$ on the interval $(c', c'']$ equals the count of conjugate points of $s = c''$ on the interval $[c', c'')$.*

It should be understood that γ is a sensed curve and that it is not reversed in sense to define the conjugate points of $s = c''$ preceding c''.

PART V

Global Theorems

Chapter 9

Simplifying concepts

§24. I, <u>extremal nondegeneracy</u>. II, <u>singleton extremals</u>. The <u>object of Part</u> I of this section is to complete the proof of Measure Theorem 11.1.

Let a point A_1 be prescribed on M_n. Recall that a finite extremal γ of J on M_n issuing from A_1 is termed <u>degenerate</u>, if the terminal point P of γ is conjugate on γ to A_1. We here admit that γ may be nonsimple. Theorem 11.1 may be restated as follows.

<u>Theorem</u> 24.1. <u>For</u> A_1 <u>prescribed on</u> M_n, <u>let</u> $((A_1))$ <u>be the set of points</u> P <u>on</u> M_n <u>such that</u> A_1 <u>is conjugate to</u> P <u>on some extremal</u> γ <u>issuing from</u> A_1. <u>Then the measure of</u> $((A_1))$ <u>on</u> M_n <u>is zero.</u>

If, for example, M_n is an n-sphere S_n with $n > 1$, then P is in $((A_1))$, if and only if P is diametrically opposite to A_1 on S_n or coincides with A_1. If M_n is a torus T, it is readily shown that $((A_1))$ is empty for some points A_1 on T, while for other points A_1 on T, $((A_1))$ has the power of the continuum.

The proof of Theorem 24.1 depends upon two lemmas.

<u>Introduction to Lemma</u> 24.1. To establish Theorem 24.1, an extension of Corollary 11.1 is needed in which the extremals $\Gamma_a = \Phi(\cdot, a)$ are not required to represent <u>simple</u> extremals on M_n.

To that end we suppose that an extremal γ_0 of J is given on M_n, issuing from A^1 on M_n with R-length $L(\gamma_0) = \rho$. The extremal γ_0 is representable, in accord with Theorem 22.2, as the "central image" $\pi \cdot \overrightarrow{o,\rho}$ of a regular tubular mapping,

$$(24.1) \qquad u \to \pi(u) : U(o,\rho) \to M_n$$

In Lemma 11.1 and Corollary 11.1 we can replace the presentation (ϕ, U) by the mapping π of $U(o,\rho)$ into M_n. A proper polar family $\underset{\sim}{\Phi}$ of extremal arcs $\Gamma_\alpha = \Phi(\cdot,\alpha)$ of J_{F^π} in $U(o,\rho)$ is then introduced, with $\overrightarrow{o,\rho}$ as the central extremal $\Gamma_{\underset{\sim}{0}}$. The proof of Corollary 11.1 carries over in a trivial manner to a proof of the following lemma.

Lemma 24.1. *If e is sufficiently small and X_e is the set of conjugate points in $U(o,\rho)$ of $s = 0$, on extremals $\Gamma_\alpha = \Phi(\cdot,\alpha)$ of J_{F^π} for which $\|\alpha\| < e$, then X_e has the measure 0 on $U(o,\rho)$.*

Lemma 24.1 is useful in proving Lemma 24.2. Lemma 24.2 will imply Theorem 24.1.

Lemma 24.2. *If A_1 is prescribed in M_n, together with an arbitrary positive number ρ, the subset of points P on extremals γ of J of R-length ρ issuing from A_1, conjugate to A_1, has the measure zero on M_n.*

The extremals γ of Lemma 24.2, issuing from A_1 with R-lengths ρ, will be denoted by $A_1\gamma$. Let $(\phi, U) \in \mathcal{D}M_n$ be such that $(\phi(\underset{\sim}{0})) = A_1$. A sufficiently short initial subarc of an extremal $A_1\gamma$ has an image in U, under ϕ^{-1}, with an R-unitary* direction r at $u = \underset{\sim}{0}$. We term r the *initial ϕ-direction* of $A_1\gamma$. If (ψ, V) is any other presentation in $\mathcal{D}M_n$ such that $\psi(\underset{\sim}{0}) = A_1$, the initial ψ-direction of $A_1\gamma$ is a contravariant image of the initial ϕ-direction of $A_1\gamma$.

*For R-unitary directions see Def. 7.3.

Proof of Lemma 24.2. Let r_0 be the initial ϕ-direction of an extremal $A_1\gamma_0$ on M_n. If $e > 0$ is sufficiently small, Lemma 24.1 implies the following. The extremals $A_1\gamma$ with initial ϕ-directions r such that $\|r-r_0\| < e$ and R-lengths at most ρ, bear a set[†] of conjugate points of A_1 with Lebesgue measure 0.

The ellipsoid $\mathcal{J}^0$ (Def. 7.3) of initial ϕ-directions r of extremals $A_1\gamma$ on M_n is compact. This fact and the conclusion of the preceding paragraph imply Lemma 24.2.

Theorem 24.1 follows from Lemma 24.2.

Definition 24.1. Nondegenerate point pairs (A_1, A_2). Let A_1 be prescribed on M_n and fixed. The point A_2 is then taken as any point on M_n not A_1, such that A_1 is not conjugate to A_2 on any extremal γ of J issuing from A_1. A_2 can be taken as any point not A_1 in the complement $C((A_1))$ of the set $((A_1))$ of Theorem 24.1. The points of $C((A_1))$ are everywhere dense on M_n. A pair (A_1, A_2) so chosen, is termed nondegenerate.

Extremals γ_0 of J, joining A_1 to A_2, are isolated in the sense of the following theorem. In this theorem we refer to the R-length $L(\gamma_0)$ of γ_0, as defined in (2.5).

Theorem 24.2. *Let (A_1, A_2) be a nondegenerate point pair on M_n, and γ_0 an extremal of J joining A_1 to A_2 on M_n. Let $(\phi, U) \in \mathcal{D}M_n$ be such that $\phi(\underset{\sim}{0}) = A_1$.*

If $\delta > 0$ is sufficiently small, there exists no extremal γ of J other than γ_0, joining A_1 to A_2 on M_n and such that

† A set which may be empty.

(24.2) $$|L(\gamma) - L(\gamma_0)| < \delta, \quad \|r - r_0\| < \delta,$$

where r *and* r_0 *are, respectively, "initial* ϕ*-directions" of* γ *and* γ_0 *at the origin in* U.

Set $L(\gamma_0) = \rho$. According to Theorem 22.2 there exists a regular tubular mapping

(24.3) $$u \to \pi(u) : U(o, \rho) \to M_n$$

such that $\gamma_0 = \pi \cdot \overrightarrow{o, \rho}$. As in §9, let $\underset{\sim}{\Phi}$ be a proper polar family of extremals in $U(o, \rho)$ of the W-integral J_{F^π}, with $\overrightarrow{o, \rho}$ as central extremal and $s = 0$ as pole. The mapping

(24.4) $$(s, a) \to \Phi(s, a) : [o, \rho] \times B_m \to U(o, \rho) \qquad [\text{cf. } (9.2)']$$

which defines the family $\underset{\sim}{\Phi}$, has a nonnull Jacobian when $s = \rho$ and $\|a\| = 0$, since $s = \rho$ is not conjugate to $s = 0$ on $\overrightarrow{o, \rho}$ by hypothesis. Hence the mapping (24.4) into $U(o, \rho)$ is biunique if the n-tuples (s, a) are restricted to a sufficiently small neighborhood N of the n-tuple $(s, a) = (\rho, \underset{\sim}{0})$.

This means that for $(s, a) \in N$, the point s on the extremal $\Phi(\cdot, a)$ coincides with the terminal point of $\overrightarrow{o, \rho}$ if and only if $(s, a) = (\rho, \underset{\sim}{0})$. Theorem 24.2 follows.

Theorem 24.2 has the following corollary.

Corollary 24.1. If (A_1, A_2) *is a nondegenerate pair of points on* M_n, *the following is true:*

(i) <u>The number (possibly 0) of extremals γ of J joining A_1 to A_2 on M_n with R-lengths $L(\gamma)$ at most a prescribed positive constant, is finite.</u>

(ii) <u>Hence the number of extremals γ of J joining A_1 to A_2 with $L(\gamma)$ unconditioned, is finite or countably infinite.</u>

The proof of this corollary is left to the reader. It is suggested that (i) is a consequence of Theorem 24.2 and (ii) a consequence of (i).

Corollary 24.1 has the following two corollaries.

<u>Corollary</u> 24.2. <u>If (A_1, A_2) is a nondegenerate pair of points on M_n the following is true.</u>

(α) <u>The number (possibly 0) of extremals γ of J joining A_1 to A_2 on M_n with J-lengths $J(\gamma)$ at most a prescribed constant B is finite.</u>

(β) <u>Hence the number of extremals of J joining A_1 to A_2 with $J(\gamma)$ unconditioned is finite or countably infinite.</u>

We shall prove (α) of Corollary 24.2 by making use of the affirmation (a_4) of §20, namely,

(a_4) If $\Delta(p, q) \leq \underset{\sim}{m}$ and $p \neq q$ then $0 < d(p, q) \leq \underset{\sim}{\kappa}$. It follows from (a_4) that an arc of an extremal γ with J-length at most $\underset{\sim}{m}$ has an R-length at most $\underset{\sim}{\kappa}$. Corresponding to B of (α) there exists an integer μ such that $B \leq \mu \underset{\sim}{m}$. Hence an extremal γ with J-length at most $\mu \underset{\sim}{m}$ has an R-length at most $\mu \underset{\sim}{\kappa}$. Corollary 24.2($\alpha$) is implied then by Corollary 24.1(i).

Corollary 24.2(β) follows from Corollary 24.2(α).

<u>Corollary</u> 24.3. <u>If (A_1, A_2) is a ND point pair no extremal length is</u>

assumed by more than a finite number of extremals joining A_1 to A_2.

Corollary 24.3 leads us to a concept which is a major concern in §24.

Definition 24.2. Singleton extremals. A J-length a, assumed by an extremal joining a ND point pair (A_1, A_2), is called singleton if there is but one extremal γ joining A_1 to A_2 with J-length a. The extremal γ is then called singleton. Part II of §24 begins here.

Objective of Part II of §24. We are concerned with the prospective proof of Theorem 26.1. Theorem 26.1 is our first theorem on extremal-homology relations. The extremals which are presupposed in Theorem 26.1 may or may not be singleton. In any case the simplest proof occurs when these extremals are singleton. Theorem 26.1, modified by the assumption that its extremals are singleton, will be called Singleton Theorem 26.1. It is our aim to verify the following lemma.

Lemma 24.3. Singleton Theorem 26.1 implies Theorem 26.1.

A proof of Lemma 24.3 requires a definition of the terms employed in Theorem 26.1. Such definitions follow.

Ordinary values relative to $J_{A_1}^{A_2}$. A value $\beta \in R$ which is not the J-length of an extremal joining A_1 to A_2 will be called ordinary relative to $J_{A_1}^{A_2}$. The value β may or may not be a J-length of an admissible curve joining A_1 to A_2.

The extremals S_β of Theorem 26.1. Let β be ordinary relative to $J_{A_1}^{A_2}$. Let S_β be a maximal set of extremals γ of J, which join a ND pair (A_1, A_2) and which are mutually A_1A_2-homotopic under* the J-level β.

---

*That is, homotopic relative to the class of broken extremals ζ joining A_1 to A_2 with $J(\zeta) < \beta$.

Definition 24.3. The type numbers of S_β. Each extremal γ of S_β is assigned an index k equal to the number of conjugate points of A_1 preceding A_2 on γ, counting each conjugate point with its multiplicity. The type number m_k of S_β is then the number of extremals in the set S_β with the index k.

Definition 24.4. The space $\bar{Z}_\beta^{(\nu)}$. In Definition 21.1 a subspace $\bar{Z}^{(\nu)}$ of the product space $(M_n)^\nu$ has been introduced. We here suppose that ν is so large that

$$\beta < \underset{\sim}{m}(\nu + 1) \qquad (\underset{\sim}{m} \text{ of Def. 19.4}) . \tag{24.5}$$

Let $\bar{Z}_\beta^{(\nu)}$ be the subspace of $\bar{Z}^{(\nu)}$ of ν-tuples $z \in \bar{Z}^{(\nu)}$ such that $\underset{\smile}{J}^{\nu}(z) \leq \beta$. Let κ be the maximum of the indices of extremals of S_β. In the special case in which $\bar{Z}_\beta^{(\nu)}$ is a pathwise connected space, Theorem 26.1 gives $\kappa + 1$ relations between the type numbers $m_0, m_1, \ldots, m_\kappa$ and the connectivities $R_0, R_1, \ldots, R_\kappa$ of $\bar{Z}_\beta^{(\nu)}$ over a field $\mathcal{K}$. The field $\mathcal{K}$ is prescribed and fixed.

The space $\bar{Z}_\beta^{(\nu)}$ may not be pathwise connected. If, for example, M_n is a torus in R^3, $\bar{Z}_\beta^{(\nu)}$ is not pathwise connected. If however, M_n is an n-sphere with $n > 1$, $\bar{Z}_\beta^{(\nu)}$ is pathwise connected. More generally, $\bar{Z}_\beta^{(\nu)}$ is pathwise connected if and only if, for ν-tuples $z \in \bar{Z}_\beta^{(\nu)}$, the broken extremals $\zeta^{\nu}(z)$ are mutually A_1A_2-homotopic under the J-level β. This leads us to the study of the pathwise connected components C of $\bar{Z}_\beta^{(\nu)}$. They are spaces $[g]_\beta^{\nu}$ called g-admissible vertex spaces.

<u>Definition</u> 24.5. <u>The component</u> $[g]_\beta^\nu$ <u>of</u> $\bar{Z}_\beta^{(\nu)}$. Let g be an extremal of minimum J-length in the set S_β. Let $[g]_\beta^\nu$ denote the pathwise component C of $\bar{Z}_\beta^{(\nu)}$ such that $g \in \zeta^\nu(C)$, or equivalently $S_\beta \subset \zeta^\nu(C)$.

g-Admissible vertex spaces $[g]_\beta^\nu$ are central in what follows.

For the purpose of proving Lemma 24.3 the following characterization of Theorem 26.1 will suffice.

(i) Theorem 26.1 gives a set of $\kappa + 1$ relations between the type numbers $m_0, m_1, \ldots, m_\kappa$ of S_β and the connectivities of $[g]_\beta^\nu$ over the field $\mathcal{K}$.

That Theorem 26.1 follows from the corresponding Singleton Theorem 26.1 is trivially implied by a lemma now to be formulated.

Replacement Lemma 24.4. Corresponding to the above set S_β of extremals of J there exists a W-integral $\hat{J}$ on the R-manifold M_n such that the following is true.

(a_1) The value β is ordinary relative to $\hat{J}_{A_1}^{A_2}$.

(a_2) The set $\hat{S}_\beta$ of extremals $\hat{\gamma}$ of $\hat{J}$ joining A_1 to A_2 with $\hat{J}(\hat{\gamma}) < \beta$, is in 1-1 correspondence with S_β.

(a_3) The extremals of $\hat{S}_\beta$ and ND and singleton.

(a_4) Corresponding extremals γ and $\hat{\gamma}$ have the same index, and are A_1A_2-homotopic both under the J-level β and the $\hat{J}$-level β.

(a_5) For a prescribed positive $e > 0$ corresponding extremals γ and $\hat{\gamma}$ have Fréchet distances less than e and J-lengths such that

$$|J(\gamma) - \hat{J}(\hat{\gamma})| < e .$$

(a_6) <u>If</u> g <u>and</u> $\hat{g}$ <u>are extremals in the sets</u> S_β <u>and</u> $\hat{S}_\beta$, <u>respectively, with minimum</u> J <u>and</u> $\hat{J}$<u>-lengths, the</u> ith <u>connectivities of</u> $[g]_\beta^\nu$ <u>and</u> $[\hat{g}]_\beta^\nu$ <u>are equal for each integer</u> i.

By virtue of (a_4) of Lemma 24.4 the type numbers m_k of S_β equal the corresponding type numbers $\hat{m}_k$ of $\hat{S}_\beta$. Moreover, the connectivities R_i and $\hat{R}_i$ of the associated spaces $[g]_\beta^\nu$ and $[\hat{g}]_\beta^\nu$ are equal, by virtue of property (a_6). Hence relations established between the entities $m_0, \ldots, m_\kappa$ and $R_0, \ldots, R_\kappa$ in a proof of Singleton Theorem 26.1 hold without change in Theorem 26.1 between identical entities $\hat{m}_0, \ldots, \hat{m}_\kappa$ and $\hat{R}_0, \ldots, \hat{R}_\kappa$.

<u>Proof of Lemma</u> 24.4. Results similar to Lemma 24.4 (without (a_6)) have been established in Morse [6] in the case when the W-integrals are R-integrals. The proof of $(a_1), \ldots, (a_5)$ is similar and will appear in a separate paper, together with a proof of (a_6). An integral $\hat{J}$ is defined depending on a parameter β. The definition will be such that $\hat{J}$ reduces to J when $\beta = 0$ and when β is positive and sufficiently small, serves as the integral $\hat{J}$ of Lemma 24.4.

§25. Theorem on Index Functions. Finite extremals* γ of J on M_n which do not afford a relative minimum to J require a special local characterization. This characterization is afforded by associating with γ a special real-valued function I_γ, termed an index function for γ. The definition of I_γ makes use of "elementary extremals" of J, as introduced in Definition 19.4. Elementary extremals were defined in a uniform manner on M_n, under the assumption that M_n is compact and that F is positive definite. These assumptions are here made.

Let (A_1, A_2) be a disjoint point pair on M_n and γ an extremal of J that joins A_1 to A_2. We suppose that $L(\gamma) = \rho$ and that R-length s is measured along γ from A_1. Let

$$0 = s_0 < s_1 < \cdots < s_\nu < s_{\nu+1} = \rho \qquad (\nu > 0) \tag{25.1}$$

be values of s on γ which divide γ into elementary extremals of J-length $< \underset{\sim}{m}$. Each index function I_γ associated with γ is defined with the aid of a product manifold $((M))_\gamma^\nu$ termed a manifold frame and characterized as follows.

Definition 25.1. A manifold frame $((M))_\gamma^\nu$. Set $m = n - 1$. A product

$$M^{(1)} \times \cdots \times M^{(\nu)} \tag{25.2†}$$

of m-dimensional manifolds on M_n, cutting across γ at the respective points,

---

*The extremals γ may be simple or nonsimple.

†Each carrier $|M^{(i)}|$ is supposed included in the range of some presentation of the set $\mathcal{B}M_n$.

(25.3) $$\gamma(s_i) = q_i \qquad (i = 1, \ldots, \nu)$$

of γ is introduced and denoted by $((M))^{\nu}_{\gamma}$. The manifolds of $((M))^{\nu}_{\gamma}$ are required to satisfy the following conditions.

<u>Condition</u> (1). For $i = 1, \ldots, \nu$, $M^{(i)}$ is given by an injective diff

(25.4) $$\alpha \to M^{(i)}(\alpha) : B_m \to M_n \qquad (m = n - 1)$$

of class* C^{∞}, where B_m is an open origin-centered m-ball in R^m and $\alpha = (\alpha^1, \ldots, \alpha^m) \in B_m$.

<u>Condition</u> (2). $M^{(i)}(\underset{\sim}{0}) = q_i \qquad (i = 1, \ldots, \nu)$.

<u>Condition</u> (3). The carriers $|M^{(i)}|$ have disjoint closures on M_n.

<u>Condition</u> (4). The manifolds $M^{(i)}$ are not tangent to γ at q_i and meet γ only in the respective points q_i.

<u>Condition</u>† (5). If $p_i \in |M^{(i)}|$ for $i = 1, \ldots, \nu$, successive points in the sequence,

(25.5) $$A_1, p_1, \ldots, p_{\nu}, A_2 ,$$

can be joined on M_n by elementary extremals of J on M_n.

<u>Definition</u> 25.2. <u>The index-function I_{γ}, based on a manifold frame $((M))^{\nu}_{\gamma}$</u>. The ν-tuples, $(p_1, \ldots, p_{\nu})$ of the sequence (25.5) will be supposed

---

* B_m admits a regular mapping of class C^{∞} into the coordinate domain U of an appropriate presentation $(\phi, U) \in \mathcal{L}M_n$ and thereby into $\phi(U)$.

† Condition (5) will be satisfied if the carriers $|M^{(i)}|$ are included in sufficiently small neighborhoods of the respective points q_i. This follows from Lemma 20.1. Conditions (1) to (4) are satisfied without difficulty. We term γ the <u>central extremal</u> of $((M))^{\nu}_{\gamma}$.

represented as follows. Let $B_{m\nu}$ be the ν-fold product $(B_m)^\nu$ in $R^{m\nu}$. Let $\underset{\sim}{v}$ be an arbitrary $m\nu$-tuple,

$$(25.6)' \qquad \underset{\sim}{v} = (v^1, v^2, \ldots, v^{m\nu}) \qquad (m = n - 1)$$

in $B_{m\nu}$, of the more specific form,

$$(25.6)'' \qquad (a_1^1, \ldots, a_1^m; \ldots; a_\nu^1, \ldots, a_\nu^m) \ .$$

Corresponding to each $m\nu$-tuple $\underset{\sim}{v} \in B_{m\nu}$ of form (25.6)' and an equivalent $m\nu$-tuple (25.6)'', a ν-tuple of vertices,

$$(25.7) \quad p_1, \ldots, p_\nu = M^{(1)}(a_1^1, \ldots, a_1^m), \ldots, M^{(\nu)}(a_\nu^1, \ldots, a_\nu^m) = ((M))_\gamma^\nu(\underset{\sim}{v})$$

is defined. We shall refer to the corresponding injective diff

$$(25.7)' \qquad \underset{\sim}{v} \to ((M))_\gamma^\nu(\underset{\sim}{v}) : (B_m)^\nu \to Z^{(\nu)} \ . \qquad \text{(See Def. 21.1 of } Z^{(\nu)}.)$$

Let $\hat{E}(\underset{\sim}{v})$ be the broken extremal on M_n defined by the sequence of $\nu + 1$ elementary extremals of J which join the successive points (25.5), subject to (25.7). Note that $\hat{E}(\underset{\sim}{0}) = \gamma$. For $\underset{\sim}{v} \in B_{m\nu}$ we set

$$(25.8) \qquad I_\gamma(\underset{\sim}{v}) = J(\hat{E}(\underset{\sim}{v})) \ ,$$

thereby defining an <u>index function</u> I_γ based, as we shall say, on the <u>manifold frame</u> $((M))_\gamma^\nu$.

It follows from Theorem 20.1 that the index function I_γ is of class C^∞ on its domain $B_{m\nu}$. To verify this one writes the integral on the right of

(25.8) as the sum of the ν values of the integral J, taken over the respective elementary extremals of $\hat{E}(\underset{\sim}{v})$. One then applies Theorem 20.1 to the i-th of these integrals and infers thereby that I_γ is of class C^∞.

A first lemma follows.

Lemma 25.1. For an index function I_γ, the $m\nu$-tuple $\underset{\sim}{v} = \underset{\sim}{0}$ is a critical point of I_γ defining the extremal $\hat{E}(\underset{\sim}{0}) = \gamma$.

Any two successive elementary extremals of $\hat{E}(\underset{\sim}{0})$ form a subarc γ_0 of γ, that is, a simple subcurve γ_0. This is in accord with Definition 19.4 of an elementary extremal. In accord with this definition γ_0 is locally representable by an extremal arc $\underset{\sim}{z}$ in the coordinate domain U of a "canonical" presentation (ϕ, U) with pole $\phi(\underset{\sim}{0}) = q$ and associated integral J_F. See $(\underset{\sim}{A})$ of §19. Holding the endpoints of $\underset{\sim}{z}$ fast, the classical first variation of J_F based on $\underset{\sim}{z}$ vanishes. If then v^j is a coordinate of $\underset{\sim}{v}$ it follows that

$$\frac{\partial}{\partial v^j} J(\hat{E}(\underset{\sim}{v}))\Big|^{\underset{\sim}{v}=\underset{\sim}{0}} = \underset{\sim}{0} \,. \tag{25.9}$$

Hence Lemma 25.1 is true.

Definition 25.3. A subframe of $((M))^\nu_\gamma$. A manifold frame $((\hat{M}))^\nu_\gamma$ whose ν component manifolds $\hat{M}^{(i)}$ have carriers

$$|\hat{M}^{(i)}| \subset |M^{(i)}| \qquad (i = 1, \ldots, \nu) \tag{25.10}$$

will be called a subframe of $((M))^\nu_\gamma$.

We state a lemma which involves this definition.

Lemma 25.2. *If* I_γ *and* $\hat{I}_\gamma$ *are index functions of the extremal* γ, *based respectively, on the manifold frame* $((M))_\gamma^\nu$ *and a subframe* $((\hat{M}))_\gamma^\nu$ *of* $((M))_\gamma^\nu$, *the index and nullity of* $\underset{\sim}{v} = \underset{\sim}{0}$ *as a critical point of* I_γ *equals the index and nullity of* $\underset{\sim}{v} = \underset{\sim}{0}$ *as a critical point of* $\hat{I}_\gamma$.

This lemma is immediate if one recalls the definition of the index and nullity of a critical point of I_γ or $\hat{I}_\gamma$.

Consider I_γ first. For $i, j = 1, \ldots, m\nu$, set

$$a_{ij} = \frac{\partial^2}{\partial v^i \partial v^j} I_\gamma(\underset{\sim}{v})\Big|^{\underset{\sim}{v}=\underset{\sim}{0}} . \tag{25.11}$$

The *nullity* of $\underset{\sim}{v} = \underset{\sim}{0}$ as a critical point of I_γ is, by definition, the nullity of the matrix $\|a_{ij}\|$. The *index* of $\underset{\sim}{v} = \underset{\sim}{0}$ is, by definition, the index of the quadratic form with $\|a_{ij}\|$ as its matrix of coefficients. The case of $\hat{I}_\gamma$ is similar.

Under the hypotheses of the lemma there exists a C^∞-diff $\underset{\sim}{v} \to \Theta(\underset{\sim}{v})$ of a neighborhood N of the origin in $R^{m\nu}$, onto a neighborhood of the origin in $R^{m\nu}$, leaving the origin fixed and such that for $\underset{\sim}{v} \in N$

$$\hat{I}_\gamma(\underset{\sim}{v}) = I_\gamma(\Theta(\underset{\sim}{v})) . \tag{25.12}$$

The lemma follows.

The function I_γ defined in (25.8) is called an *index function for* γ because of the validity of the following basic theorem.

Index Theorem 25.1. *Let* (A_1, A_2) *be a disjoint point pair on* M_n,

γ an extremal of J, joining A_1 to A_2 and $((M))_\gamma^\nu$ a "frame" of m-manifolds $M^{(i)}$ cutting across γ. If I_γ is the index function "based" on $((M))_\gamma^\nu$, then the following is true.

(a) The nullity of the critical point $\underset{\sim}{v} = \underset{\sim}{0}$ of I_γ equals the multiplicity of A_2 as a conjugate point of A_1 on γ.

(b) The index of the critical point $\underset{\sim}{v} = \underset{\sim}{0}$ is the count* of conjugate points of A_1 on γ preceding A_2.

Theorem 25.1 is a consequence of Theorem 17.2 of Morse [1], as we shall see.

Set $\rho = L(\gamma)$. We represent γ in accord with Theorem 22.2 as the central image $\pi \cdot \overrightarrow{o, \rho}$ of a regular tubular mapping

$$u \to \pi(u) : U(o, \rho) \to M_n \qquad \text{(see §22)} \tag{25.13}$$

where $U(o, \rho)$ is an open neighborhood in R^n of the arc $\overrightarrow{o, \rho}$ of the u^1-axis. The points $q_1, \ldots, q_\nu$ of (25.3) on γ will be images under π of the point on the u^1-axis at which u^1 has the respective values,

$$s_1, \ldots, s_\nu \qquad \text{(of (25.1))} . \tag{25.14}$$

Further conditions on the mapping π. Let $[o, \rho]_e$ be the set of points in R^n within a distance e of the arc $\overrightarrow{o, \rho}$. One can take U as the set $[o, \rho]_e$,

* This count is, by definition, the sum of the multiplicities of conjugate points A_1 preceding A_2. It will be called the index of γ.

if e is sufficiently small. Let

$$\Pi_1, \ldots, \Pi_\nu \qquad \text{(cf. (13.1) of Morse [1])} \tag{25.15}$$

be open planar m-balls of radius e orthogonal to the u^1-axis at the respective points $u^1 = s_i$ of (25.1). Rectangular coordinates in R^n are here denoted by $u^1, \ldots, u^n$. The m-balls (25.15) admit Monge representations in terms of the coordinates $u^2, \ldots, u^n$ as parameters. Set

$$(u^2, \ldots, u^n) = (a^1, \ldots, a^m) \qquad (m = n-1)\ . \tag{25.16}$$

None of the ν-manifolds $M^{(i)}$ is tangent to γ at its point of intersection with γ. The m-balls Π_i are similarly nontangent to the u^1-axis. It is for this reason that the following lemma can be established, as the reader will readily verify.

Lemma 25.3. *If $e > 0$ is sufficiently small, $[o,\rho]_e$ is the domain of a tubular mapping π which maps $\overrightarrow{o,\rho}$ onto γ and the respective m-balls*

$$\Pi_1, \ldots, \Pi_\nu \tag{25.17}$$

diffeomorphically onto manifolds

$$\hat{M}^{(1)}, \ldots, \hat{M}^{(\nu)} \tag{25.18}$$

which define a subframe of $((M))^\nu_\gamma$. Specifically π shall map a point of Π_i whose coordinates $(u^1, \ldots, u^n)$ in R^n are

$$(s_i, a^1, \ldots, a^m) \qquad (\|a\| < e) \tag{25.19}$$

into the point $\hat{M}^{(i)}(a)$ of $\hat{M}^{(i)}$.

Completion of the proof of Theorem 25.1. We seek the index and nullity of the critical point $\underset{\sim}{v} = \underset{\sim}{0}$ of the index function I_γ based on a frame $((M))_\gamma^\nu$ prescribed for γ. According to Lemma 25.2 the desired index and nullity are the index and nullity of $\underset{\sim}{v} = \underset{\sim}{0}$ as a critical point of the index function $\hat{I}_\gamma$ based on the subframe $((\hat{M}))_\gamma^\nu$ of $((M))_\gamma^\nu$. We suppose the manifolds $\hat{M}^{(i)}$ of $((\hat{M}))_\gamma^\nu$ related to the m-balls Π_i as in Lemma 25.3.

A revaluation (25.22) of $\hat{I}(\underset{\sim}{v})$. For the regular tubular mapping π of Lemma 25.3 let the W-preintegrand F^π be defined as in §23. According to Corollary 23.1, the arc $\overrightarrow{o,\rho}$ in $[o,\rho]_e$ is an extremal of J_{F^π}. Let A' and A'' be, respectively, the endpoints* of the arc $\overrightarrow{o,\rho}$. We suppose that the $m\nu$-tuple $\underset{\sim}{v}$ is represented as in (25.6)'' and, for $i = 1, \ldots, \nu$, let $P_i(\underset{\sim}{v})$ then be the point in Π_i with coordinates $(s_i, a_i^1, \ldots, a_i^m)$. Successive points in the sequence,

$$A', P_1(\underset{\sim}{v}), \ldots, P_\nu(\underset{\sim}{v}), A'', \tag{25.20}$$ †

can be joined in $[o,\rho]_e$ by extremal arcs

$$z_0(\underset{\sim}{v}), z_1(\underset{\sim}{v}), \ldots, z_\nu(\underset{\sim}{v}) \tag{25.21}$$

of J_{F^π} whose images, under π, are the successive elementary extremals on M_n of the broken extremal $\hat{E}(\underset{\sim}{v})$ of (25.8). It follows from (25.8) that

$$\hat{I}_\gamma(\underset{\sim}{v}) = \sum_{i=0}^{\nu} J_{F^\pi}(z_i(\underset{\sim}{v})) \qquad (\|\underset{\sim}{v}\| < e). \tag{25.22}$$

---

* Note that $\pi(A')$ and $\pi(A'')$ are the endpoints A_1 and A_2 of γ.

† The sequence (25.20) is of the nature of the sequence $\underset{\sim}{P}(v)$ of (13.13), Morse [1].

<u>A final evaluation of $\hat{I}_\gamma(\underset{\sim}{v})$</u>. In order to apply Theorem 17.2 of Morse [1], $\hat{I}_\gamma(\underset{\sim}{v})$ must be evaluated in terms of the nonparametric theory. To that end let the Euler preintegrand f, with values $f(x, y, p)$, be defined by F^π, as in (8.3). If e is sufficiently small, the extremals (25.21) are "Monge curves" in $[o, \rho]_e$ whose x-parameterized "mates" will be respectively denoted by $h_0(\underset{\sim}{v}), \ldots, h_\nu(\underset{\sim}{v})$. It follows from (25.22) and Theorem 8.1 that

$$\hat{I}_\gamma(\underset{\sim}{v}) = \sum_{i=0}^{\nu} J_f(h_i(\underset{\sim}{v})) \qquad (\|\underset{\sim}{v}\| < e) . \tag{25.23}$$

<u>An application of Theorem 17.2 of Morse [1]</u>. From the definition (14.6) on page* 86 of Morse [1], and from (25.23), we see that $\hat{I}_\gamma$ is an "index function," in the sense of Morse [1], based on the extremal $g = \overrightarrow{o, \rho}$ of f. In Morse [1], $\lambda = 0$ and g is parameterized by $x = u^1 = s$. With this understood, Theorem† 17.2 of Morse [1] implies Theorem 25.1 by virtue of a simple extension of Corollary 10.2: a point on the extremal γ of J with parameter $s > 0$ is a conjugate point of A_1 with a multiplicity μ, if and only if the point on $g = \overrightarrow{o, \rho}$ with coordinate $u^1 = s$ is a conjugate point with multiplicity μ of the initial point of g.

Statements (a) and (b) of Theorem 25.1 now follow, respectively, from statements (i) and (ii) of Theorem 17.2 of Morse [1].

<u>Interpretation of Theorem 25.1</u>. Set $\mu = m\nu$, where $m = n - 1$. The definition of the index function

---

*In the case at hand $\lambda = 0$, $\Theta(a) \equiv 0$; $X^1(a) \equiv 0$, $X^2(a) \equiv \rho$ in Def. 14.1 of Morse [1].

†In the sense of Morse [1], the index and nullity of g are, by definition, the index and nullity of $\underset{\sim}{v} = \underset{\sim}{0}$ as a critical point of an "index function" of g.

$$\underset{\sim}{v} \to I_\gamma(\underset{\sim}{v}) : (B_m)^\nu \to R$$

culminates in (25.8). The μ-tuple $\underset{\sim}{v}$ is represented as in (25.6)', or equivalently as in (25.6)''. Moreover, $I_\gamma(\underset{\sim}{0}) = J(\gamma)$ and the point $\underset{\sim}{v} = \underset{\sim}{0}$ in R^μ is a critical point of I_γ in accord with Lemma 25.1. If γ is ND, the critical point $\underset{\sim}{v} = \underset{\sim}{0}$ of I_γ is ND according to (a) of Theorem 25.1.

Unless otherwise stated we now assume that (A_1, A_2) is a ND point pair. Under this hypothesis γ is ND. We seek a Taylor's representation of $I_\gamma(\underset{\sim}{v}) - I_\gamma(\underset{\sim}{0})$ near the origin. Our representation of $I_\gamma(\underset{\sim}{v}) - I_\gamma(\underset{\sim}{0})$ will aid in characterizing the topological changes in the sublevel sets of I_γ as the level increases or decreases through the critical value $I_\gamma(\underset{\sim}{0})$.

<u>Introduction to Lemma</u> 25.4. For $\underset{\sim}{v}$ restricted to a sufficiently small open neighborhood N of the origin in R^μ, Taylor's formula, with a remainder in integral form, gives an identity,

$$I_\gamma(\underset{\sim}{v}) - I_\gamma(\underset{\sim}{0}) \equiv a_{ij}(\underset{\sim}{v})v^i v^j \qquad (\underset{\sim}{v} \in N) \tag{25.24}$$

where the coefficients (Jordan [1], p. 249)

$$a_{ij}(\underset{\sim}{v}) = \int_0^1 (1-t) \frac{\partial^2}{\partial v^i \partial v^j} I_\gamma(tv^1, \ldots, tv^\mu)\, dt$$

are of class C^∞ on N. Moreover, the μ-square determinant $|a_{ij}(\underset{\sim}{0})| \neq 0$, since $\underset{\sim}{0}$ is a ND critical point of I_γ. The identity (25.24) leads to a much more revealing identity.

Let D_e be an origin-centered closed μ-ball of radius e in R^μ. Let $(x_1, \ldots, x_\mu)$ be rectangular coordinates of a point $\underset{\sim}{x}$ in R^μ.

Lemma 25.4. *Suppose that the origin* $\underset{\sim}{0}$ *in* R^μ *is a* ND *critical point of* I_γ *of index* k. *If then* D_e *is an origin-centered* μ-*ball in* R^μ *with a sufficiently small radius* e, *there exists a diff,*

$$(25.25) \qquad \underset{\sim}{x} \to \psi(\underset{\sim}{x}) : D_e \to R^\mu \qquad (\mu = m\nu)$$

of class C^∞ *onto a neighborhood of the origin in* R^μ *such that* $\psi(\underset{\sim}{0}) = \underset{\sim}{0}$ *and*

$$(25.26) \qquad I_\gamma(\psi(\underset{\sim}{x})) - I_\gamma(\underset{\sim}{0}) \equiv -x_1^2 - \cdots - x_k^2 + x_{k+1}^2 + \cdots + x_\mu^2 \qquad (\underset{\sim}{x} \in D_e).$$

Were the coefficients $a_{ij}(\underset{\sim}{v})$ in (25.24) constants and the determinant $|a_{ij}| \neq 0$, the diff ψ could be taken as a nonsingular homogeneous linear transformation, in accord with Lagrange's method of reducing a quadratic form to a sum of signed squares. See Bôcher [1], p. 131. In the case at hand one can replace the Lagrange transformations by similarly defined diffs, provided D_e is sufficiently small in radius. For further details see Morse [5], p. 45.

A second form for Lemma 25.4. Corollary 25.1 gives Lemma 25.4 a form which will be used in proving our first global theorem in §26.

Our first global theorem is concerned with the ensemble of extremals which join A_1 *to* A_2, *which have a prescribed homotopy type and* J-*lengths*

<u>less than a prescribed constant</u> β.

Such extremals can be considered as EB-extremals with nonsingular* vertex sets $\underset{\sim}{q} = (q_1, \ldots, q_\nu)$. The integer ν is taken so large that

(25.27) $$(\nu+1)\underset{\sim}{m} > \beta \qquad \text{(See Def. 19.4 for } \underset{\sim}{m}.)$$

and then fixed. According to Theorem 21.1 each such vertex set $\underset{\sim}{q}$ is a critical ν-tuple of the vertex mapping

$$(p_1, \ldots, p_\nu) \to \mathcal{J}^\nu(\underset{\sim}{p}) : \check{Z}^{(\nu)} \to R .$$

In Lemma 25.4 there is given a typical extremal γ joining A_1 to A_2. A nonsingular set of vertices subdividing γ is given by the set

$$(q_1, \ldots, q_\nu) = (\gamma(s_1), \ldots, \gamma(s_\nu))$$

of (25.3). $\mathcal{J}^\nu$ is the function whose critical points are to be studied. We wish to give Lemma 25.4 a form in which $\mathcal{J}^\nu$ replaces I_γ in (25.26). This is to conform to the conditions of the general theory as formulated in Morse and Landis [1].

It follows from the definition of $I_\gamma(\underset{\sim}{v})$ in (25.8) and Definition 21.0 of $\mathcal{J}^{(\nu)}$ that

(25.28) $$I_\gamma(\underset{\sim}{v}) \equiv \mathcal{J}^\nu(\underset{\sim}{p}) \qquad (\underset{\sim}{v} \in (B_m)^\nu)$$

---

* The vertex set $(q_1, \ldots, q_\nu)$ is <u>nonsingular</u> in that no two successive vertices coincide. Cf. §21.

subject to the condition $p = ((M))^{\nu}_{\gamma}(v)$ of (25.7). By virtue of (25.25) and (25.28),

$$(25.29) \qquad I_{\gamma}(\psi(x)) \equiv \mathcal{J}^{\nu}(\Phi_{\gamma}(x)) \qquad (x \in D_e)$$

where $\Phi_{\gamma}(x)$ is a v-tuple $(p_1, \ldots, p_{\nu})$

$$(25.30) \qquad \Phi_{\gamma}(x) = ((M))^{\nu}_{\gamma}(\psi(x)) \in Z^{(\nu)} \qquad (x \in D_e).$$

Such ν-tuples serve to define an <u>injective</u> diff

$$(25.31) \qquad x \to \Phi_{\gamma}(x) : D_e \to Z^{(\nu)}$$

of D_e into $((M))^{\nu}_{\gamma}$. The dimension of D_e is $m\nu = (n-1)\nu$, that of $Z^{(\nu)}$ is $n\nu$. With this understood, Lemma 25.4 implies the following.

<u>Corollary</u> 25.1. <u>Let</u> γ <u>be an</u> ND <u>extremal of index</u> k <u>joining</u> A_1 <u>to</u> A_2. <u>Let</u> $q = (q_1, \ldots, q_{\nu})$ <u>be the nonsingular vertex set defined by</u> (25.3). <u>Set</u> $\mu = m\nu$. <u>If</u> D_e <u>is a sufficiently small origin-centered closed</u> μ-<u>ball in</u> R^{μ} <u>there exists an injective</u> *diff* Φ_{γ} *: (25.30)* <u>of</u> D_e <u>into</u> $((M))^{\nu}_{\gamma}$ <u>such that</u> $\Phi_{\gamma}(0) = q$ <u>and</u>

$$(25.32) \quad \mathcal{J}^{\nu}(\Phi_{\gamma}(x)) - \mathcal{J}^{\nu}(q) \equiv -x_1^2 - \cdots - x_k^2 + x_{k+1}^2 + \cdots + x_{\mu}^2 \qquad (x \in D_e).$$

Subsequent sections will make extensive use of deformations of ν-tuples in $Z^{(\nu)}$. In this context the following notion is essential.

<u>Definition</u> 25.4. <u>Broken-extremal projections into</u> $((M))^{\nu}_{\gamma}$. Let

$\underset{\sim}{q} = (q_1, \ldots, q_\nu)$ be the nonsingular vertex set defined by (25.3). Let $z = (z_1, \ldots, z_\nu)$ be a ν-tuple in an open neighborhood N of $\underset{\sim}{q}$ relative to $Z^{(\nu)}$. If N is sufficiently small the broken extremal $\zeta^\nu(z)$ joining A_1 to A_2 will meet the (n-1)-dimensional manifolds $M^{(i)}$ of $((M))^\nu_\gamma$ in unique successive points

$$\pi_1(z), \ldots, \pi_\nu(z) \tag{25.33}$$

(some possibly identical) which vary continuously with $z \in N$ and define a ν-tuple $\pi(z)$ in $((M))^\nu_\gamma$. We term $\pi(z)$ the <u>broken extremal projection</u> of z <u>into</u> $((M))^\nu_\gamma$. Note that $\pi(z) = z$, if z is in $((M))^\nu_\gamma$. As will be seen in Appendix IV, ν-tuples z sufficiently near $\underset{\sim}{q}$ can be deformed locally in $Z^{(\nu)}$ into their projections $\pi(z)$.

Chapter 10

Reduction to Critical Point Theory

§26. Extremal homology relations under a finite J-level. In Chapter 10 the study of extremal homology relations will be restricted to the study of extremals joining an ND point pair (A_1, A_2). Theorem 26.1 concerns the set S_β of extremals of J characterized in Part II of §24. Theorem 26.1 is a consequence of Theorem 7.1 of Landis-Morse [1]. The paper of Landis-Morse was written specially for this application. A principal hypothesis of the Landis-Morse Theorem is formulated in terms of a special deformation called a traction. We shall define a traction in a background of deformation theory.

Deformations. Let t be a real variable termed the time. Let $I = [0, 1]$ denote an interval for t. With us a deformation D of a subspace A of a topological space χ is a continuous mapping

$$(26.0) \qquad (p, t) \to D(p, t) : A \times I \to \chi$$

such that $D(p, 0) \equiv p$ for $p \in A$. Let F be a continuous mapping of χ into R. The deformation D will be called an F-deformation if

$$(26.1) \qquad F(D(p, 0)) \geq F(D(p, t)) \qquad \big(((p, t) \in A \times I)\big) .$$

For p fixed in A, the partial mapping

$$t \to D(p, t) : I \to \chi$$

is called the <u>trajectory</u> of p under D. More generally if Y is a subspace of A, the restriction of the mapping D to $Y \times I$ is well-defined. We say that D deforms Y <u>on</u> the set $D(Y, I)$ <u>onto</u> a <u>final image</u> $D(Y, 1)$. If $\chi \supset B \supset D(Y, 1)$ we say that D deforms Y <u>on</u> $D(Y, I)$ <u>into</u> B. The point $D(p, t)$ will be termed the <u>replacement</u> of p at the time t and denoted by p^t.

<u>Retracting Deformations</u>. A deformation D of A onto a set $B \subset A$ is termed a <u>deformation retracting</u> A onto B, if D deforms A on A onto B and if $D(p, t) = p$ for $p \in B$ and $0 \leq t \leq 1$. See Borsuk [1].

Retracting deformations will be supplemented by a larger class of deformations termed <u>tractions</u>.

<u>Definition</u> 26.1. <u>Tractions</u>. A deformation D of a subspace A of χ will be termed a <u>traction of</u> A <u>into a subspace</u> B, if D deforms A <u>on</u> A <u>into</u> B and deforms B <u>on</u> B.

Tractions imply homology relations. We refer to the singular homology theory introduced by Eilenberg [1]. Homology groups will be taken over an arbitrary commutative field $\mathcal{K}$. The q-th homology group of a subspace A of a topological space χ will be denoted by $H_q(A)$.

Each deformation retracting A onto B is a traction of A into B, but a traction of A into B is not, in general, a deformation retracting, A onto B, as examples easily show. However, a traction of A into B and a deformation retracting A onto B share a fundamental property: they imply the existence of an isomorphic mapping of $H_q(A)$ onto $H_q(B)$ for each q. This theorem motivated our definition of a traction but will not be directly used in this section. See Landis-Morse [1].

Introduction to Theorem 7.1 of Landis-Morse. Theorem 7.1 of Landis-Morse concerns a real-valued function F defined and continuous on a compact metric space X. Corresponding to each value $b \in R$ we shall set

$$F_b = \{p \in X \mid F(p) \leq b\} \tag{26.2}$$

and term F_b a sublevel set of X. A set F_b may be empty. We suppose that $X = F_\beta$ for some $\beta \in R$.

A traction induced critical point, σ of F, will now be defined. To define such points we shall make use of an origin-centered open μ-ball D_e^μ in R^μ of radius e, of points

$$\underset{\sim}{x} = (x_1, \ldots, x_\mu) \in R^\mu . \tag{26.3}$$

Definition 26.2. A traction induced critical point σ. Let k be an integer on the range $0, 1, \ldots, \mu$. Let $d \leq \beta$ be given. A point $\sigma \in F_d$ with $F(\sigma) = a < d$ will be called a critical point of F, induced by an F-traction T of F_d, and of index k, if the following two conditions are satisfied.

Local Condition I. For some sufficiently small positive e there exists an injective homeomorphism,

$$\underset{\sim}{x} \to \Phi_\sigma(x) : D_e^\mu \to F_d , \tag{26.4}$$

onto a topological μ-ball Λ_σ with $\sigma = \Phi_\sigma(\underset{\sim}{0})$ and

$$(26.5) \qquad F(\Phi_\sigma(x)) - a \equiv -x_1^2 - \cdots - x_k^2 + x_{k+1}^2 + \cdots + x_\mu^2 \qquad (\underset{\sim}{x} \in D_e^\mu) .$$

Global Condition II. For some value $c \in R$ such that $a > c > a - e^2$, there exists an F-traction of F_d into $\Lambda_\sigma \cup F_c$.

When a is the absolute minimum of F, F_c is empty and Condition II reduces to the condition that there exist an F-traction of F_d into Λ_σ.

The topological μ-ball Λ_σ is not required to be a neighborhood of σ relative to F_d. It contains σ and is termed a μ-ball of index k induced by the traction T of F_d into $\Lambda_\sigma \cup F_c$.

Notation for Theorem 7.1 of Landis-Morse [1]. There are given r values

$$(26.6) \qquad a_1 < a_2 < \cdots < a_r \qquad (a_i < \beta)$$

of F of which a_1 is the absolute minimum of F. Points

$$(26.7) \qquad \sigma_1, \sigma_2, \ldots, \sigma_r \qquad (\sigma_i \in \check{X})$$

are given such that for $i = 1, 2, \ldots, r$, $F(\sigma_1) = a_i$. We shall condition the points σ_i and values

$$(26.8) \qquad c_0 < c_1 < \cdots < c_{r-1} < c_r$$

requiring first that

$$(26.9) \qquad c_0 < a_1 < c_1 < a_2 < \cdots < a_r < c_r \qquad (c_r = \beta)$$

Hypothesis of Landis-Morse Theorem 7.1. For some sufficiently large integer μ, for suitable choices of the constants (26.8) such that (26.9) holds, and for $i = 1, \dots, r$, σ_i shall be a critical point of F of index k_i induced by an F-traction T_i of F_{c_i} into $\Lambda_{\sigma_i} \cup F_{c_{i-1}}$ in accord with Definition 26.2. Note that T_r is an F-traction of X into $\Lambda_r \cap F_{c_{r-1}}$ and that T_1 is an F-traction of F_{c_1} into Λ_1.

Landis-Morse Theorem 7.1. Let κ be the maximum of the indices k_i of the traction induced critical points $\sigma_1, \dots, \sigma_r$ of F and $m_0, m_1, \dots, m_\kappa$ the corresponding type numbers. If R_q is the q-th connectivity of X over the field $\mathcal{K}$, then R_q is finite and vanishes for $q > \kappa$. Moreover

$$
\begin{aligned}
m_0 &\geq R_0 \\
m_1 - m_0 &\geq R_1 - R_0 \\
m_2 - m_1 + m_0 &\geq R_2 - R_1 + R_0 \qquad (26.10)\\
- \quad - \quad - &\geq - \quad - \quad - \\
m_\kappa - m_{\kappa-1} + - \dots (-1)^\kappa m_0 &= R_\kappa - R_{\kappa-1} + - \cdots (-1)^\kappa R_0
\end{aligned}
$$

The relations (26.10) imply that $m_q \geq R_q$ for each $q \geq 0$.

The reader is referred to Landis-Morse [1] for proof.

The data for Theorem 26.1. As indicated in §24, there is given an ND point pair (A_1, A_2) a number β, ordinary relative to $J_{A_1}^{A_2}$ and a maximal set

$$S_\beta = (\gamma_0, \dots, \gamma_r) \qquad (J(\gamma_i) < \beta) \tag{26.11}$$

of extremals of J which join A_1 to A_2 and are mutually A_1A_2-homotopic under the J-level β. If g is an extremal of the set S_β with minimum J-length, $[g]_\beta^\nu$ denotes the pathwise component of C of $\bar{Z}_\beta^{(\nu)}$ such that $\zeta^\nu(C)$ contains g and hence includes S_β. See Def. 24.5.

Theorem 26.1. *Let κ be the maximum of the indices k_i of the extremals of S_β and let $m_0, \ldots, m_\kappa$ be the type numbers* (Def. 24.3) *of S_β. Let R_q denote the q-th connectivity over the field $\mathcal{K}$ of the space $[g]_\beta^\nu$ of ν-tuples. Then each R_q is finite, $R_0 = 1$, $R_q = 0$ for $q > \kappa$ and the relations of the Landis-Morse Theorem hold between the type numbers $m_0, \ldots, m_\kappa$ and the connectivities $R_0, \ldots, R_\kappa$.*

Proof of Theorem 26.1. As has been shown in §24, to prove Theorem 26.1 it suffices to prove the corresponding Singleton Theorem 26.1, that is Theorem 26.1 in the case in which each of the extremals $\gamma_0, \ldots, \gamma_r$ of S_β is singleton.

Singleton Theorem 26.1 will be shown to follow from the Landis-Morse Theorem 7.1. To that end, for ν-tuples $z \in [g]_\beta^\nu$ set

(26.12) $$\mathcal{J}^\nu(z) = f^\nu(z)$$

where $\mathcal{J}^\nu$ is the mapping introduced in §21. The mapping

(26.13) $$z \to f^\nu(z) : [g]_\beta^\nu \to R$$

will be identified with the mapping F of Landis-Morse. The domains of both F and f^ν are compact metric spaces. The domain $[g]_\beta^\nu$ is also pathwise connected, so that $R_0 = 1$. Let

(26.14)' $$b_0 < b_1 < \cdots < b_r$$

be the J-lengths of the respective extremals $\gamma_0, \ldots, \gamma_r$. Let

(26.14)'' $$\tau_0, \ldots, \tau_r$$

be the J-normal ν-tuples which partition the extremals $\gamma_0, \ldots, \gamma_r$ in accord with Definition 21.5. Singleton Theorem 26.1 will follow from Landis-Morse Theorem 7.1 if we verify the following.

(a) The ν-tuples $\tau_0, \ldots, \tau_r$ are traction induced critical points of f^ν, in the sense in which the points (26.7) are traction induced critical points of F. The index of τ_i as critical point equals the index of γ_i as extremal.

Proof of (a). The meaning of (a) must be made clear.

As a submanifold of $(M_n)^\nu$, $[g]^\nu_\beta$ has a dimension $n\nu$. As an $n\nu$-tuple in local coordinates of $[g]^\nu_\beta$, each τ_i is a critical point of f^ν in the ordinary sense, in accord with Theorem 21.1. Constants

(26.15)[†] $$c_0 < c_1 < \cdots < c_{r-1} < c_r$$

such that

(26.16) $$b_0 < c_0 < b_1 < c_1 < \cdots < c_{r-1} < b_r < c_r \qquad (c_r = \beta)$$

will be further conditioned. Set $\mu = (n-1)_\nu$. Given $c_r = \beta$ we shall show that

[†]The needs of future papers have lead to a departure from the enumeration (26.8) of the c_i's given in Landis-Morse.

constants $c_{r-1}, c_{r-2}, \ldots, c_0$ can be chosen in the order written so that (26.16) holds and the following holds in accord with Corollary 25.1.

Lemma 26.1. *If, for* $i = 0, 1, \ldots, r$, *an origin-centered open* μ*-ball* $D^\mu_{e_i}$ *in* R^μ *has a sufficiently small radius* e_i, *there exists a diffeomorphism*

$$(26.17) \qquad x \to \Phi_i(x) : D^\mu_{e_i} \to [g]^\nu_{c_i} \qquad (\text{with } \tau_i = \Phi_i(\underset{\sim}{0}))$$

onto a topological μ*-ball,*

$$(26.18) \qquad \lambda_i = \Phi_i(D^\mu_{e_i}) \qquad (i = 0, 1, \ldots, r)$$

which is an open neighborhood of τ_i *relative to the manifold frame* $((M))^\nu_{\gamma_i}$ *and is such that*

$$(26.19) \qquad f^\nu(\Phi_i(x)) - b_i = -x_1^2 - \cdots - x_{k_i}^2 + x_{k_{i+1}}^2 + \cdots + x_\mu^2 \qquad (x \in D^\mu_{e_i})$$

where k_i *is the index of the extremal* γ_i.

Definition 26.3. *The* μ*-balls* λ_i. The μ-*balls* λ_i are well-defined by (26.18) for arbitrarily small values of e_i.

When f^ν replaces F in Definition 26.2, τ_i and λ_i can replace σ and Λ_σ, respectively, in Definition 26.2 and will satisfy the 'Local Condition' in Definition 26.2 by virtue of Lemma 26.1. The Global Condition in Definition 26.2, that τ_i be a traction induced critical point of f^ν with the index of γ_i, is satisfied. This will be shown in Appendix IV.

The Landis-Morse Theorem accordingly implies Singleton Theorem 26.1.

According to §24, Theorem 26.1 is then true even if some of its extremals fail to be singleton.

More explicitly, for $i = 1, 2, \ldots, r$, the theorem of Appendix IV that shows that τ_i is a traction induced critical point of f^ν in the singleton case refers to the set λ_i containing τ_i defined in (26.18) and to the ordered alternating set of critical values b_j of f^ν and noncritical values c_j of (26.16). It reads as follows.

Traction Theorem Ω_i, $i = 1, 2, \ldots, r$. *If* c_i *is given as in* (26.16) *then for any sufficiently small* $e_i > 0$ *and for some* $c_{i-1} \in (b_{i-1}, b_i)$ *there exists an* f^ν*-traction* T_i *of* $f^\nu_{c_i}$ *into* $\lambda_i \cup f^\nu_{c_i - 1}$.

A special traction induces the minimizing critical point τ_0 in accord with the following special theorem, proved in Appendix IV.

Traction Theorem Ω_0. *If* $c_0 \in (b_0, b_1)$ *as in* (26.16), *then for any sufficiently small* $e_0 > 0$ *there exists an* f^ν*-traction* T_0 *of* $f^\nu_{c_0}$ *into* λ_0 *of* (26.18).

§27. Fréchet numbers $\underset{\sim}{R}_i$ and related global theorems. Theorem 26.1 gives extremal homology relations subject to two conditions:

I. The extremals have J-lengths less than a prescribed constant β.

II. The extremals join a nondegenerate point pair $A_1 \neq A_2$ and are mutually A_1A_2-homotopic under the J-level β.

Theorem 27.1 below extends Theorem 26.1, removing Condition I that the extremals be bounded in J-length. It retains the condition that these extremals join a ND point pair $A_1 \neq A_2$.

Theorem 27.3 relaxes both Conditions I and II. Both of these theorems use a new set of topological invariants $\underset{\sim}{R}_i$ which we call Fréchet numbers. The range of i is 0, 1, 2, ... These numbers are the connectivities over Q of the pathwise components of a metric space $\mathcal{F}_{A_1}^{A_2}$ which we now define. See Morse [7] and Landis-Morse [2].

Fréchet spaces $\mathcal{F}_{A_1}^{A_2}$ on a manifold N_n. A compact, connected differentiable manifold N_n of class C^∞ is given. To define a distance between two curves on N_n, a metric is required on N_n. Any metric compatible with the topology of N_n will suffice. Curves on N_n are required to be proper, that is to be continuous mappings on N_n of a closed interval such that the image of no subinterval is a point.

If h is a curve on N_n, joining $A_1 \neq A_2$ on N_n, $\underset{\sim}{h}$ shall denote the class (Fréchet [1]) of all curves obtained from h by a 1-1 continuous sense-preserving change of parameter. Let hk denote the Fréchet distance (see Landis-Morse [2]) between any two curves h and k joining A_1 to A_2. If

$\underset{\sim}{h}$ and $\underset{\sim}{k}$ are the corresponding Fréchet curve classes, Frechet defines a distance $\underset{\sim\sim}{hk}$ by setting $\underset{\sim\sim}{hk} = hk$. It is immediate that $hk = kh$ and the corresponding triangle axiom holds. However, the condition $hk = 0$ does not imply that $h = k$. Fortunately, it can be shown that the distance $\underset{\sim\sim}{hk}$ satisfies all three metric axioms (Morse [3]).

Let $\mathcal{F}_{A_1}^{A_2}$ denote the resultant metric space of Fréchet classes $\underset{\sim}{h}$ of curves h joining A_1 to A_2 on N_n.

<u>The class</u> $((N_n))$. The metric space $\mathcal{F}_{A_1}^{A_2}$ is pathwise connected when N_n is an n-sphere. In general $\mathcal{F}_{A_1}^{A_2}$ is not pathwise connected. It has at most a countably infinite set of pathwise connected components. The following lemma prepares for the definition of the numbers $\underset{\sim}{R}_i$. In this lemma we shall refer to the class $((N_n))$ of all differentiable manifolds M_n of class C^∞, homeomorphic to N_n. M_n may not be diffeomorphic to N_n, as Milnor's exotic 7-spheres shows. See Milnor [1].

<u>Lemma</u> 27.1. <u>Let</u> M_n <u>be any manifold in the class</u> $((N_n))$. <u>Let</u> $A_1 \neq A_2$ <u>be an arbitrary point pair on</u> M_n <u>and</u> $\mathcal{F}_{A_1}^{A_2}$ <u>the corresponding metric Fréchet space. Pathwise components</u> $K_{A_1}^{A_2}$ <u>of</u> $\mathcal{F}_{A_1}^{A_2}$ <u>are then homeomorphic for every choice of</u> $M_n \in ((N_n))$ <u>and of a disjoint point pair</u> (A_1, A_2) <u>on</u> M_n (Landis-Morse [2]).

<u>Definition of</u> $\underset{\sim}{R}_i$, $i = 0, 1, 2, \ldots$. The connectivity $\underset{\sim}{R}_i$ over the field Q of rational numbers of the pathwise components of a Fréchet space $\mathcal{F}_{A_1}^{A_2}$ is independent of the point pair $A_1 \neq A_2$ on M_n and will be called the i-th <u>Fréchet number</u> of M_n.

The Fréchet numbers $\underset{\sim}{R}_i$ of an n-sphere S_n, $n > 1$, equal 1 if $i \equiv 0 \bmod n-1$ and vanish for other values of i. The Fréchet numbers for a projective n-plane P_n equal those of an n-sphere S_n. However, the number of components of a Fréchet space $\mathcal{J}_{A_1}^{A_2}$ is 1 for S_n and 2 for P_n $|(n = 2)$.

The Fréchet numbers for the product of μ n-spheres or projective n-planes is easily found when $n > 2$. For example, $S_3 \times S_3$ has the Fréchet numbers 1, 0, 2, 0, 3, ... For a product $N_n \times S_1$ the Fréchet numbers are those of N_n. For a manifold whose sectional Riemann curvatures are nonpositive, $\underset{\sim}{R}_0 = 1$, while $\underset{\sim}{R}_i = 0$ for $i > 0$. Lemma 27.1 implies that manifolds $M_n \in ((N_n))$ have the Fréchet numbers of N_n. That all Fréchet numbers are finite, may be true, but has not been proved. The Fréchet numbers of a surface are uniquely determined by its Euler characteristic, but the converse is not true. See page 29 of Massey [1] for Euler characteristics.

The principal theorem follows. The differentiable manifold M_n is supposed in the class $((N_n))$. On M_n an integral J is defined as in the preceding sections.

__Global Theorem__ 27.1. *Let m_r^g be the number (possibly infinite) of g-admissible extremals of index r joining a ND point pair $A_1 \neq A_2$ on M_n.*

If the type numbers m_r^g are finite, then $m_r^g > \underset{\sim}{R}_r$ for each r and the following relations are true:

$$
\begin{aligned}
m_0^g &\geq \underset{\sim}{R}_0 \\
m_1^g - m_0^g &\geq \underset{\sim}{R}_1 - \underset{\sim}{R}_0 \\
m_2^g - m_1^g + m_0^g &\geq \underset{\sim}{R}_2 - \underset{\sim}{R}_1 + R_0 \\
\ldots \qquad\qquad \ldots &\geq \ldots \qquad\qquad \ldots
\end{aligned}
$$

Comments on Theorem 27.1. The type numbers m_r^g introduced in Theorem 27.1 are not finite a priori. In fact, the author has proved the following. If the Fréchet numbers of N_n are finite, then in each class $((N_n))$ there exist differentiable manifolds M_n with R-length such that for the corresponding geodesics some of the type numbers are infinite, provided A_1, A_2 and an A_1A_2-homotopy class are properly prescribed. In any case $m_r^g \geq \underset{\sim}{R}_r$, if (A_1, A_2) is a ND point pair.

Finite type numbers. The numbers m_r^g in Theorem 27.1 are finite if M_n is the product of any finite set of spheres or projective planes of variable dimensions and if J is taken as ordinary length on such a product. The numbers m_r^g are easily computed in this case. The following definition leads to a very general condition on the Weierstrass integral J on M_n sufficient that the type numbers m_r^g of Theorem 27.1 be finite whatever the choice of the ND pair A_1, A_2 and associated A_1A_2-homotopy class.

Definition 27.1. Integrals J on M_n which are conjugate point bounded. We say that J is conjugate point bounded on M_n if there is a positive constant B such that the J-length along each extremal γ of J from its initial point p to a first conjugate point q of p (if q exists) is less than B.

We state a theorem.

Theorem 27.2. *For an integral* J *on* M_n *which is conjugate point bounded, each of the type numbers* m_r^g *is finite in Theorem* 27.1, *whatever the choice of the* ND *point pair* A_1, A_2 *and associated* A_1A_2*-homotopy class.*

A theorem similar to Theorem 27.2 is sometimes quoted in geodesic theory under the hypothesis that the sectional 2-dimensional Riemann curvatures of M_n are positive and bounded from zero. Cf. Rauch [1]. Such a theorem is implied by Theorem 27.2, but the converse is not true.

An open question. The Fréchet number $\underset{\sim}{R}_i$ is finite if m_i^g of Theorem 27.1 is finite, but the converse is not true. Whether or not the Fréchet numbers are always finite is an open question. If the Fréchet numbers are finite for any differentiable manifold N_n then, by Lemma 27.1, they are finite for each manifold in $((N_n))$. For example, they are finite for each n-sphere. They are accordingly finite for each exotic sphere of Milnor type. They are finite for a "flat" torus and hence are finite for the classical torus. They are finite and easily computable for every surface. See Cairns-Morse [1].

Degenerate point pairs. The restriction to ND point pairs in Theorem 27.1 will now be relaxed.

A general point pair PQ *on* M_n. We require only that $P \neq Q$. With the pair PQ and M_n there is given an arbitrary curve h which joins P to Q on M_n. We seek extremals of J which join P to Q on M_n in the PQ-homotopy class of h. We do not exclude extremals γ of J which retrace subarcs of γ several times.

Theorem 27.3. *Let an integral* J *be given on* M_n *which is conjugate point bounded. Let* R_k *be a Fréchet number of* M_n *which is positive. Corresponding to an arbitrary curve* h *joining* P *to* Q *there exists at least one extremal of* J *which joins* P *to* Q *in the PQ-homotopy class of* h *and bears at least* k *and at most* $k + n - 1$ *points conjugate to* P.

The theorems of this section will be proved separately. A paper entitled *Conjugate points on a limiting extremal* will be presently published in the Proceedings of the National Academy of Sciences and contain a general theorem of which Theorem 27.3 is a corollary.

§28. Introduction to proof of Global Theorem 27.1. The proof of this theorem will be presented in two memoirs with the respective titles

Memoir I. Nondegenerate point pairs in global variational analysis.

Memoir II. Fréchet numbers in global variational analysis.

Memoir I makes no use of Fréchet numbers. It is primarily concerned with analyzing the "g-admissible vertex spaces" $[g]_\beta^\nu$ of Definition 24.5. Memoir I*gives a more explicit characterization of g-admissible vertex spaces as follows.

Definition 28.1. A homotopically minimizing extremal g. A ND point pair $A_1 \neq A_2$ on M_n and a curve h joining A_1 to A_2 are prescribed. By Theorem 21.2, A_1 can be joined to A_2 by an extremal g which is A_1A_2-homotopic to h and affords a minimum to J relative to all piecewise regular curves which join A_1 to A_2 and are A_1A_2-homotopic to g. The extremal g is now held fast. An extremal which joins A_1 to A_2 and is A_1A_2-homotopic to g is called g-admissible.

Let β be any value in R such that $J(g) < \beta$ and β is J-ordinary, that is, not the J-length of any g-admissible extremal. If ν is a positive integer such that

$$J(g) < \beta < \underset{\sim}{m}(\nu+1) \qquad (\underset{\sim}{m} \text{ of Def. } 19.4) \tag{28.1}$$

g can be partitioned into $(\nu+1)$-successive elementary extremal arcs of equal J-length $< \underset{\sim}{m}$.

---

* To appear in Jour. of Diff. Geometry.

Definition 28.1. A g-admissible vertex space $[g]_\beta^\nu$. If (28.1) holds with β J-ordinary, a maximal pathwise connected subspace of the product $(M_n)^\nu$, which satisfies the following three conditions is called a g-admissible vertex space $[g]_\beta^\nu$.

Condition I. Each ν-tuple $z = (z_1, \ldots, z_\nu)$ of $[g]_\beta^\nu$ shall be such that successive points of M_n in the sequence

$$A_1, z_1, \ldots, z_\nu, A_2 \tag{28.2}$$

which are distinct can be joined by elementary extremals of J.

Condition II. The broken extremal $\zeta^\nu(z)$, joining A_1 to A_2 and defined by the successive elementary extremals joining successive points in (28.2), has a J-length $\leq \beta$.

Condition III. $[g]_\beta^\nu$ contains the ν-tuple $(z_1, \ldots, z_\nu)$ which partitions g into $\nu+1$ elementary extremals of equal J-length.

It is of prime importance to know how the connectivities of a vertex space $[g]_\beta^\nu$ vary with admissible variation of ν and β. An essential lemma affirms that if μ is an integer $> \nu$, then the homology groups of $[g]_\beta^\nu$ and $[g]_\beta^\mu$ are isomorphic.

In Theorem 27.1, m_i^g denotes the number (possibly countably infinite) of g-admissible extremals joining the ND point pair $A_1 \neq A_2$ with an index i. The principal hypothesis of Theorem 27.1 is that the numbers m_i^g (termed type numbers) are finite for each i. Lemmas of Memoir I imply the following.

Lemma 28.1. *If the type numbers* m_i^g *are finite for each integer* $i \geq 0$, *then corresponding to each such* i *there exists a number* $B_i \geq J(g)$ *with the following property.*

The ith connectivity, over the field $\mathcal{K}$, *of any g-admissible vertex space* $[g]_\beta^\nu$ *is an integer* L_i^g *independent of J-ordinary values* $\beta > B_i$ *and integers* ν *such that* $\beta < \underset{\sim}{m}(\nu+1)$.

The principal theorem of Memoir I is the following:

Theorem 28.1. *If the type numbers* m_i^g *are finite for each integer* $i \geq 0$, *Theorem 27.1 takes a valid form if the Fréchet numbers* $\underset{\sim}{R}_i$ *therein are replaced by the respective numbers* L_i^g.

Theorem 28.1 is a consequence of Lemma 28.1 and Theorem 26.1, as is shown in Memoir I.

Memoir II establishes the following. When the type numbers m_i^g are finite for each $i \geq 0$, then for each such i

$$L_i^g = \underset{\sim}{R}_i \qquad (28.3)$$

Theorem 28.1 and (28.3) imply Theorem 27.1.

The relation (28.3) is established for each choice of the ND point pair (A_1, A_2) and of a homotopically minimizing extremal g joining A_1 to A_2. To prove (28.3) one is led to the chapter of singular algebraic topology which affirms that a continuous map ϕ of one topological space χ into another χ' "induces chain transformations" of $S(\chi)$ into $S(\chi')$. In the notation of Eilenberg [1] $S(\chi)$ and $S(\chi')$ are the complexes of singular cells with carriers on χ and χ' respectively. See Morse and Cairns [1], §26.

The spaces χ and χ' which are here involved are the vertex spaces $[g]^{\nu}_{\beta}$ and the pathwise components $\mathcal{F}_g$ of the Fréchet spaces $\mathcal{F}^{A_2}_{A_1}$ of §27. An essential simplification is afforded by Theorem 1.1 of Landis-Morse [2] according to which the pathwise components $\mathcal{F}_g$ of $\mathcal{F}^{A_2}_{A_1}$ are homeomorphic and remain homeomorphic if the point pair $A_1 \neq A_2$ is replaced by any other pair of distinct points on M_n.

<u>The relevant spaces</u> χ <u>and</u> χ'. Fortunately one can prescribe an integer i in (28.3) and prove (28.3) separately for each i. One chooses a vertex space $[g]^{\nu}_{b}$ for which b is a J-ordinary value $> B_i$ of Lemma 28.1. If z is a ν-tuple in $[g]^{\nu}_{b}$, the broken extremal $\zeta^{\nu}(z)$ is well-defined, as is the Fréchet curve class, say $\underset{\sim}{\zeta}^{\nu}(z)$, of $\zeta^{\nu}(z)$. The mapping

$$z \to \underset{\sim}{\zeta}^{\nu}(z) : [g]^{\nu}_{b} \to F(g) \tag{28.4}$$

is readily shown to be continuous. The chain transformations induced by the mapping (28.4) are essential in proving (28.3) for the given i.

Mappings of subspaces of $\mathcal{F}_g$ into the above vertex space $[g]^{\nu}_{b}$ which are relevant are much more difficult to come by. Elements of $\mathcal{F}_g$ are Fréchet curve classes $\underset{\sim}{h}$ of curves h joining A_1 to A_2 and A_1A_2-homotopic to g.

One seeks mappings

$$\underset{\sim}{h} \to z(\underset{\sim}{h}) : \mathcal{F}_g \to [g]^{\nu}_{b} \tag{28.5}$$

in which $z(\underset{\sim}{h})$ is a ν-tuple in $[g]_b^\nu$ whose vertices suitably partition some curve $h \in \underset{\sim}{h}$. In general such a mapping cannot be continuous because ν is a finite fixed integer. Fortunately it is sufficient for our purposes to show that a mapping of form (28.5) can be defined and proved continuous on a special compact subset Y_i of $\mathcal{F}_g$ determined by the integer i prescribed in (28.3). That this is possible is a consequence of a "Parameterization Theorem" for Fréchet spaces established for this purpose in Morse [3].

For the given i one proves first that $L_i^g \geq \underset{\sim}{R}_i$ and then that $L_i^g \leq \underset{\sim}{R}_i$ thereby establishing (28.3). The above chain transformations and suitably defined tractions are our principal tools.

Appendix I

The existence of regular tubular mappings. Theorem 22.2 will be established in this section. Use will be made of the notation of §22.

We are concerned with the regular curve ζ : (22.1) of class C^∞ on M_n. If s_0 is given in $[a, b]$ and if $[\alpha, \beta]$ is a sufficiently short subinterval of $[a, b]$ such that s_0 equals α or β, classical implicit function theorems suffice to show that there exists a "simple" tubular* mapping

$$u \to \hat{\pi}(u) : U(\alpha, \beta) \to M_n \tag{1.1}$$

such that for each point $(c, 0, \dots, 0) \in U(\alpha,\beta) \cap |a, b|$, $\hat{\pi}(c, 0, \dots, 0) = \zeta(c)$. Such a mapping will be termed ζ-axial.

From this lemma and the compactness of the interval $[a, b]$ the following can be inferred. The interval $[a, b]$ of Theorem 22.2 can be decomposed into a finite sequence of intervals (22.3) with the respective open tubular neighborhoods,

$$V(c_1, c_2), V(c_2, c_3), \dots, V(c_\mu, c_{\mu+1}) \qquad (\mu > 1) \tag{1.2}$$

such that there exist "simple" tubular mappings,

$$u \to \pi_i(u) : V(c_i, c_{i+1}) \to M_n \qquad (i = 1, \dots, \mu) \tag{1.3}$$

each of which is "ζ-axial."

---

* In Appendix I all diffs are supposed "extendable" in the sense of Definition 22.3.

Theorem 22.2 will follow from the following lemma.

Lemma 1. *It is possible to modify the* μ *mappings* π_i *of* (1.3) *so as to obtain similarly conditioned mappings,*

(1.4) $$u \to \hat{\pi}_i(u) : U(c_i, c_{i+1}) \to M_n \qquad (i = 1, \ldots, \mu)$$

whose successive domains intersect in and only in disjoint open neighborhoods,

(1.5) $$X_2, X_3, \ldots, X_\mu$$

of the respective points

(1.6) $$(c_2, 0, \ldots, 0), (c_3, 0, \ldots, 0), \ldots, c_\mu(c_\mu, 0, \ldots, 0)$$

on the u^1*-axis and are such that*

(1.7) $$\hat{\pi}_{i-1}(u) \equiv \hat{\pi}_i(u) \qquad (u \in X_i;\ i = 2, 3, \ldots, \mu) .$$

The proof of this lemma in the special case in which $\mu = 2$ will make clear the mode of proof in the general case.

When $\mu = 2$ we are concerned with the mappings π_1 and π_2. The point $(c_2, 0, \ldots, 0)$ of the u^1-axis is in both $V(c_1, c_2)$ and $V(c_2, c_3)$ and is mapped by π_1 and π_2 into $\zeta(c_2)$. The mappings π_1 and π_2 are "ζ-axial." If $e > 0$ is sufficiently small, $(c_2, 0, \ldots, 0)$ has an open e-neighborhood N_2^e in $V(c_1, c_2)$ which admits a transition diff θ_2 into $V(c_2, c_3)$ such that

(1.8) $$\pi_1(u) = \pi_2(\theta_2(u)) \qquad (u \in N_2^e)$$

under θ_2 points on the u^1-axis in N_2^e are necessarily invariant.

<u>Definition of $\hat{\pi}_1$ and $\hat{\pi}_2$.</u> Let a positive constant $\varepsilon < e$ be given with $2\varepsilon < \min(c_2-c_1, c_3-c_2)$. Set

(1.9) $$\begin{cases} U(c_1, c_2) =^* [c_1, c_2]_\varepsilon \\ U(c_2, c_3) = [c_2, c_3]_\varepsilon \end{cases}$$

subject to two additional conditions on ε. We suppose that ε is so small that

(1.10) $$U(c_1, c_2) \subset V(c_1, c_2) \;;\; U(c_2, c_3) \subset V(c_2, c_3)$$

Note that the intersection of the two domains (1.9) is $N_2^\varepsilon \subset N_2^e$. If ε is sufficiently small the restriction $\theta | N_2$ admits an extension as a C^∞-diff Θ of $U(c_2, c_3)$ into $V(c_2, c_3)$ that leaves points of the u^1-axis in $U(c_2, c_3)$ invariant. The mappings $\hat{\pi}_1$ and $\hat{\pi}_2$ are defined by setting

(1.11) $$\begin{cases} \hat{\pi}_1(u) = \pi_1(u) & (u \in U(c_1, c_2)) \\ \hat{\pi}_2(u) = \pi_2(\Theta(u)) & (u \in U(c_2, c_3)) \end{cases}$$

---

* $[c_1, c_2]_\varepsilon$ denotes the open set of points in R^n whose distances from $|c_1, c_2|$ are less than ε.

The mappings

$$(1.12)\qquad \begin{cases} u \to \hat{\pi}_1(u) : U(c_1, c_2) \to M_n \\ u \to \hat{\pi}_2(u) : U(c_2, c_3) \to M_n \end{cases}$$

thereby defined are clearly "simple" tubular mappings into M_n. They are "ζ-axial," since π_1 and π_2 are "ζ-axial."

The mappings (1.12) satisfy the condition (1.7) when $i = 2$. This is because

$$(1.13)\qquad [c_1, c_2]_\varepsilon \cap [c_2, c_3]_\varepsilon = N_2^\varepsilon \subset N_2^e$$

so that (1.7) holds by virtue of the definition (1.11) of $\hat{\pi}_1$ and $\hat{\pi}_2$.

Thus Lemma 1 is true when $\mu = 2$. When $\mu > 2$ we leave the proof of Lemma 1 to the reader.

Proof of Theorem 22.2. The domains $U(c_i, c_{i+1})$ of Definition 22.2 (iii) are here identified with the domains $U(c_i, c_{i+1})$ of Lemma 1. The domain $U(a, b)$ of Theorem 22.2 is taken as the union of the μ domains of Lemma 1. The mapping π of Theorem 22.2 is then overdefined by the μ conditions

$$\pi(u) = \hat{\pi}_i(u) \qquad\qquad (u \in U(c_i, c_{i+1}))$$

These conditions are consistent by virtue of (1.7).

Theorem 22.2 follows.

Appendix II

Minimizing extremals, phasewise near a minimizing extremal. Let F and $\mathcal{f}$ be preintegrands associated with a presentation (ϕ, U). We shall make use of the "phase space" terminology of §14.

Theorem 1. Suppose that F is positive regular on U (Definition 14.2). Let

$$(1.1) \qquad \underset{\sim}{w} : s \to w(s) : [0, s_1] \to U$$

be an extremal arc of J_F bearing no point conjugate to $s = 0$. There then exists an open neighborhood N_w of $|w|$ in U and a positive constant e_w so small that the following is true.

Any extremal arc γ of J_F such that Ph γ is in the "e_w-tube of Ph $\underset{\sim}{w}$," has a carrier in N_w and gives a proper strong minimum to J_F relative to admissible curves K joining the endpoints of γ in N_w.

The following lemma implies Theorem 1.

Lemma 1. Under the hypotheses of Theorem 1 there exists an open neighborhood N_w of $|w|$ in U such that the following is true.

If N_1 and N_2 are sufficiently small open neighborhoods in N_w of the first and last points of $\underset{\sim}{w}$, any extremal arc η of J_F in N_w which joins a point in N_1 to a point in N_2, affords a proper minimum to J_F, relative to any admissible curve in N_w which joins the endpoints of η in N_w.

Proof that Lemma 1 implies Theorem 1. Suppose that Lemma 1 holds with N_w, N_1 and N_2 properly chosen. If e_w is then a sufficiently small positive constant, any extremal γ of J_F in N_w such that $Ph\,\gamma$ is in the "e_w-tube of $Ph\,\underset{\sim}{w}$," will have an extremal extension η_γ (possibly γ) in N_w which joins a point in N_1 to a point in N_2. By Lemma 1 η_γ (and hence γ) will afford a proper minimum to J_F relative to any admissible curve which joins its endpoints in N_w.

Thus Theorem 1 is true if Lemma 1 is true.

Proof of Lemma 1. The hypotheses and conclusions of Lemma 1 are invariant under any transition diffeomorphism. Hence no generality is lost in the proof of Lemma 1 if we assume that the extremal $\underset{\sim}{w}$ is a subarc of the u^1-axis with $u^1(s) \equiv s$ on $\underset{\sim}{w}$.

Let f be the Euler preintegrand with values $f(x, y, p)$ defined by F as in (8.3). Let g be the x-parameterized "mate" (8.8) of $\underset{\sim}{w}$. Then g is an extremal of J_f. Along g, $0 \leq x \leq s_1$ and $g_\mu(x) \equiv 0$ for $\mu = 1, \ldots, m = n-1$. The extremal g of J_f is free of conjugate points, since $\underset{\sim}{w}$ is so free. Theorem 5.1 and Lemma 6.5 of Morse [1] can be applied to g. If one interprets Lemma 6.5 in accord with its proof, the following is implied.

Lemma 2. Under the hypotheses of Lemma 1, if N_w, N_1 and N_2 of Lemma 1 are sufficiently small, an arbitrary point in N_1 can be joined to an arbitrary point in N_2 by a unique extremal η in N_w such that the following is true.

The extremal η is a proper subarc of an extremal in a field Γ_η of *extremals of* J_F *defined by a proper polar family of extremals in* U *whose pole is exterior to* N_w *and whose carrier includes* N_w.

Completion of proof of Lemma 1. The fields Γ_η are Mayer fields Γ_0 in the sense of §17. If one identifies the extremal η with the extremal $\underset{\sim}{z}$ of Theorem 18.1, the proof of Theorem 18.1 shows that η affords a proper strong minimum to J_F, relative to admissible curves which join the end-points of η on the carrier $|\Gamma_\eta|$ of Γ_η and, in particular, on N_w.

Thus Lemma 1 is true and implies Theorem 1.

Theorem 1 has an essential corollary.

Corollary 1. *Let* (ϕ, U), F, $\underset{\sim}{w}$, N_w *and* e_w *be given as in Theorem* 1. *Let* γ_1 *and* γ_2 *be two extremal arcs of* J_F *such that* Ph γ_1 *and* Ph γ_2 *are in the "*e_w*-tube of* Ph $\underset{\sim}{w}$". *If then* γ_1 *and* γ_2 *issue from a point* Q_1 *in* N_w *with different initial directions,* γ_1 *and* γ_2 *cannot intersect in a second point* Q_2.

Suppose, on the contrary, that γ_1 and γ_2 intersect in a second point Q_2. Let $\overline{\gamma_1}$ and $\overline{\gamma_2}$ be the subarcs of γ_1 and γ_2, respectively, which join Q_1 to Q_2. Then by Theorem 1 $J_F(\overline{\gamma_1})$ is both greater and less than $J_F(\overline{\gamma_2})$. Hence γ_1 and γ_2 cannot intersect in a second point.

Thus Corollary 1 is true.

1.

Appendix III

The differentiable product manifold $(M_n)^\nu$. If $|M_n|$ denotes the Hausdorff carrier of the differentiable manifold M_n, the ν-fold product $|M_n|^\nu$ is a Hausdorff manifold of dimension $n\nu$ on which a differentiable manifold $(M_n)^\nu$ will now be defined. A Riemannian structure will be given to $(M_n)^\nu$.

A sequence

$$(1) \qquad (\theta_1, Y_1), (\theta_2, Y_2), \ldots, (\theta_\nu, Y_\nu)$$

of presentations in $\mathcal{L}\, M_n$ will serve to define a presentation $(\Theta, \underset{\sim}{Y})$ of an open subset of $|M_n|^\nu$. The subset

$$(2) \qquad \underset{\sim}{Y} = Y_1 \times Y_2 \times \cdots \times Y_\nu$$

of $R^{n\nu}$ is well-defined. For $k = 1, \ldots, \nu$ let y_k be a point $(y_k^1, \ldots, y_k^n) \in Y_k$. We introduce the $n\nu$-tuple

$$(3) \qquad \underset{\sim}{y} = (y_1, \ldots, y_\nu) = (y_1^1, \ldots, y_1^n; \ldots; y_\nu^1, \ldots, y_\nu^n)$$

of coordinates and set

$$(4) \qquad \Theta(\underset{\sim}{y}) = (\theta_1(y_1), \ldots, \theta_\nu(y_\nu)) .$$

The mapping

(5) $$\underset{\sim}{y} \to \Theta(\underset{\sim}{y}) : \underset{\sim}{Y} \to |M_n|^\nu$$

is a homeomorphism of $\underset{\sim}{Y}$ onto an open subset $\Theta(\underset{\sim}{Y})$ of $|M_n|^\nu$ to be termed a product presentation $(\Theta, \underset{\sim}{Y})$ of $|M_n|^\nu$.

Product presentations of $|M_n|^\nu$ are readily seen to be pairwise compatible in the sense of Morse-Cairns [1], page 31. The union of their ranges is $|M_n|^\nu$. The set $\mathcal{A}$ of all presentations of open subsets of $|M_n|^\nu$ which are C^∞-compatible with the product presentations of $|M_n|^\nu$, is a maximal set of pairwise C^∞-compatible presentations of open subsets of $|M_n|^\nu$. This maximal set of presentations is denoted by $\mathcal{L}(M_n)^\nu$. The manifold $|M_n|^\nu$, taken with the set $\mathcal{L}(M_n)^\nu$ of presentations of $|M_n|^\nu$ is, by definition, the differentiable manifold $(M_n)^\nu$. It is of class C^∞.

A Riemannian metric for $(M_n)^\nu$. Let k have the range $1, \ldots, \nu$ and i, j the range $1, \ldots, n$. Let* (θ_k, Y_k) be the kth presentation in $\mathcal{L} M_n$ entered in (1). With (θ_k, Y_k) and with each point $y_k \in Y_k$, there is associated a positive definite quadratic form

(6) $$(ds_k)^2 = a^k_{ij}(y_k)dy^i_k dy^j_k \qquad \text{(k not summed)}$$

This is the square of the differential of arc length on the open subset* $\theta_k(Y_k)$ of $|M_n|$, with local coordinates $y^1_k, \ldots, y^n_k$. Given the product presentation $(\Theta, \underset{\sim}{Y})$ as in (4) and (5), the $n\nu$-tuple $\underset{\sim}{y}$ of (3) gives local coordinates on $\Theta(\underset{\sim}{Y})$. The square $(ds)^2$ of the differential of arc length in terms of these local

---

* There is no summation as to k.

coordinates is (by definition) the positive definite quadratic form

$$(7) \qquad (ds)^2 = a^k_{ij}(y_k)dy^i_k dy^j_k$$

where a summation is understood with respect to k, as well as to i, j, over the respective ranges of k and i, j.

Let K be the class of product presentations in $\mathcal{L}(M_n)^\nu$. The ranges of the presentations in K cover $|M_n|^\nu$. With each presentation in K, an R-preintegrand has been associated by the definition (7). It is seen that these R-preintegrands are pairwise compatible, since the R-preintegrands f associated with the respective presentations in $\mathcal{L}M_n$ are compatible.

It remains to associate R-preintegrands with the presentations in $\mathcal{L}(M_n)^\nu$ not in K. That this can be done so that the resultant set of R-preintegrands is composed of pairwise compatible R-preintegrands, follows from Theorem 19.7 of Morse-Cairns.

In Appendix IV a major use of products of deformations will be made in the sense of the following definition.

Definition I. The product deformation $\hat{D}D$. There is given a deformation D of a subspace A of a topological space χ into a subspace B of χ. There is also given a deformation $\hat{D}$ of a subspace $\hat{A}$ of χ into a subspace $\hat{B}$ of χ. If $\hat{A} \supset B$, a product deformation $\hat{D}D = \Pi$ of A into $\hat{B}$ is defined as follows.

The parameter t for Π shall again vary from 0 to 1. Given a point $p \in A$ let p^1 be the final image of p under D. Apart from parameterization the trajectory of p under Π will be the trajectory of p under D, followed by the trajectory of p^1 and $\hat{D}$. More precisely, for $p \in A$ we set

$$\Pi(p, t) \equiv D(p, 2t) \qquad (0 \leq t \leq \frac{1}{2})$$

(8)

$$\Pi(p, t) \equiv \hat{D}(p^1, 2t-1) \qquad (\frac{1}{2} \leq t \leq 1)$$

thereby defining Π. Suppose that a third deformation Δ on χ of $A^* \subset \chi$ into $B^* \subset \chi$ is given. If $A^* \supset \hat{B}$ a deformation $\Delta(\hat{D}D)$ of A into B^* is well-defined as $\Delta\Pi$.

1.

Appendix IV

The existence of the tractions T_i of §26. The object of Appendix IV is to complete the proof of Theorem 26.1 by proving the existence of the f^ν-tractions,

(1) $$T_0, T_1, \ldots, T_r ,$$

characterized at the end of §26. We begin with T_r and the associated ν-tuple τ_r in the list

(2) $$\tau_0, \tau_1, \ldots, \tau_r \qquad \text{(see (26.14)'')}$$

of J-normal ν-tuples which partition the respective extremals $\gamma_0, \ldots, \gamma_r$ of (26.11).

By virtue of Definition 26.2 the proof that T_r exists will complete the proof that the ν-tuple τ_r is a traction induced critical point of the mapping

(3) $$f^\nu = \mathcal{J}^\nu | [g]^\nu_\beta \qquad (\beta = c_r \text{ of } (26.16)).$$

The domain $[g]^\nu_\beta$ of f^ν is denoted by f^ν_β.

In §26 we have set $\mu = (n-1)\nu$ and introduced the topological μ-ball

(4) $$\lambda_r = \Phi_r(D^\mu_{e_r}) \qquad \text{(of (26.18))}$$

as an open neighborhood of τ_r relative to $((M))^\nu_{\gamma_r}$. Here Φ_r is a diff

(26.17) for which (26.19) holds identically. For the extremal γ_r see (26.11).

<u>Traction Theorem</u>[†] Ω_r. <u>If</u> e_r <u>is a sufficiently small positive constant, then for some value</u> $c \in (b_{r-1}, b_r)$ <u>there exists an</u> f^ν<u>-traction</u> T_r <u>of</u> f^ν_β <u>into</u> $\lambda_r \cup f^\nu_c$. <u>Here</u> $b_{r-1} = f^\nu(\tau_{r-1})$ <u>and</u> $b_r = f^\nu(\tau_r)$.

The proof of this theorem is naturally presented in two parts. In <u>Part</u> I we prove a simpler theorem which does not involve the concept of a traction or the index of the extremal γ_r and in which λ_r is replaced by an $n\nu$-dimensional neighborhood of τ_r relative to f^ν_β. In <u>Part</u> II we return to an $(n-1)\nu$-dimensional neighborhood of τ_r relative to the manifold frame $((M))^\nu_{\gamma_r}$. Definition 26.2 of a traction induced critical point determines our procedures.

Appendix IV, Part I

We shall state the principal theorem of Part I. In this theorem and elsewhere we shall refer to open e-neighborhoods of compact subsets of the product $(M_n)^\nu$, defining such neighborhoods as the ensemble of points on $(M_n)^\nu$ less than an R-distance e from the given subset. In particular, use is made of 2e-neighborhoods $(\tau_r)_{2e}$ of τ_r. We begin with the case $r > 0$.

<u>Theorem</u> 1. <u>If</u> $e > 0$ <u>is sufficiently small, then for some value</u> $c \in (b_{r-1}, b_r)$ <u>there exists an</u> f^ν<u>-deformation</u> θ_e <u>of</u> f^ν_β <u>on itself which leaves</u> ν<u>-tuples in</u> $(\tau_r)_e$ <u>fixed and deforms</u> f^ν_β <u>into</u>

$$(\tau_r)_{2e} \cup f^\nu_c . \tag{5}$$

[†]We suppose first that $r > 0$. We shall return to the case $r = 0$.

θ_e is not known to be a traction of f^ν_β into the set (5), since θ_e is not known to deform the set (5) on itself. However, the condition that the ν-tuples in $(\tau_r)_e$ be fixed is an aid in proving that the Traction Theorem is valid.

To proceed with the proof of Theorem 1, special notation is required. The constant e appears first in Theorem 1.

The sets U^τ and U^τ_e. The ensemble of critical ν-tuples $\tau_0, \ldots, \tau_r$ will be denoted by U^τ. U^τ_e shall denote the e-neighborhood of U^τ. U^τ_{2e} is similarly defined.

Conditions K on e. We suppose that e is so small that the open 2e-neighborhoods

$$(6) \qquad (\tau_0)_{2e}, \ldots, (\tau_r)_{2e}$$

are disjoint and included in the interior of f^ν_β and that, for $i \neq j$, neighborhoods $(\tau_i)_{2e}$ and $(\tau_j)_{2e}$ contain no ν-tuples at a common f^ν-level. Other conditions on the smallness of e will be added.

The J-normal subset ω of f^ν_β. Let ω denote the set of J-normal ν-tuples in f^ν_β. Cf. Def. 21.5. The set ω is compact and contains the set U^τ. The symbol $\underset{\sim}{C}$ preceding a subset of f^ν_β, such as ω, will denote the complement of the set relative to f^ν_β.

Ordinary ν-tuples of f^ν_β. A ν-tuple $z \in f^\nu_\beta$ will be termed ordinary if $\zeta^\nu(z)$ is not an extremal.

f^ν-deformations G_e and Δ^η_e of f^ν_β. Theorem 1 will be proved with the aid of two f^ν-deformations denoted by G_e and Δ^η_e, respectively. Deformations such as these will be said to deform a ν-tuple $z \in f^{\nu}_{\beta}$ effectively if $f^{\nu}(z) > f^{\nu}(z^1)$, where $z^1 = G_e(z, 1)$ is the final image of z under the deformation. The following two lemmas will be proved after G_e and Δ^{η}_e have been defined.

Lemma 1. *For e conditioned as above there exists an f^{ν}-deformation G_e of f^{ν}_{β} on itself which leaves ν-tuples in $C\ell\, U^{\tau}_e$ fixed. Among the ν-tuples which G_e effectively deforms is any ν-tuple z in $\underset{\sim}{C}\, U^{\tau}_{2e}$ whose final image $G_e(z, 1)$ is not in ω.*

Lemma 2. *For e conditioned as above and for $\eta > 0$ sufficiently small, there exists an f^{ν}-deformation Δ^{η}_e of f^{ν}_{β} on itself which leaves ν-tuples in U^{τ}_e fixed and effectively deforms each ν-tuples $z \in \omega$ not in $C\ell\, U^{\tau}_e$.*

Theorem 1 will be proved with the aid of the product deformation $\Delta^{\eta}_e G_e$ of f^{ν}_{β}.

Definition of the f^{ν}-deformation G_e. Given a ν-tuple $\underset{\sim}{p} = (p_1, \ldots, p_{\nu})$ in f^{ν}_{β}, let $\zeta^{\nu}(\underset{\sim}{p})$ be the broken extremal of §21 that joins A_1 to A_2. Let

$$z(\underset{\sim}{p}) = (z_1(\underset{\sim}{p}), \ldots, z_{\nu}(\underset{\sim}{p})) \tag{7}$$

be the ν-tuple of successive points on $\zeta^{\nu}(\underset{\sim}{p})$ which divide $\zeta^{\nu}(\underset{\sim}{p})$ into $\nu + 1$ successive subcurves of equal J-length. The condition $\beta < \underset{\sim}{m}(\nu+1)$ on ν of (24.5) implies that this J-length is less than $\underset{\sim}{m}$ of Definition 19.4.

To define G_e we shall first define an auxiliary f^{ν}-deformation G.

Under G, the i-th vertex of the replacement $\underset{\sim}{p}^t$ of $\underset{\sim}{p}$ at the time t shall move along $\zeta^{\nu}(\underset{\sim}{p})$ from p_i to $z_i(p)$ as t increases from 0 to 1. Let p^t_i be the vertex which replaces p_i at the time t. The movement of

p_i^t shall be such that the J-length of the subcurve of $\zeta^\nu(\underset{\sim}{p})$ from A_1 to p_i^t changes at a constant rate. G is thus the deformation defined by the mapping

$$(8) \qquad (\underset{\sim}{p}, t) \to \underset{\sim}{p}^t : f_\beta^\nu \times [0, 1] \to f_\beta^\nu .$$

That G, as defined by (8), is continuous follows from the results of §20. That G is an f^ν-deformation follows from the minimizing properties of the elementary extremals defined by the successive vertices of $\underset{\sim}{p}^t$.

G_e shall differ from G in that ν-tuples in U_e^τ shall remain fixed under G_e. Under G_e, ν-tuples in $\underset{\sim}{C}U_{2e}^\tau$ shall be deformed as under G. For $0 \leq t' \leq 1$ ν-tuples $z \in U_{2e}^\tau$ at an R-distance $e + \hat{t}e$ from U^τ shall be deformed under G_e as under G until the time $t = \hat{t}$ is reached and held fast thereafter.

<u>Proof of Lemma</u> 1. Lemma 1 is clearly true, possibly excepting its affirmation concerning final images $z^1 \notin \omega$ under G_e of ν-tuples $z \in \underset{\sim}{C}U_{2e}^\tau$.

If $z^1 \notin \omega$, successive elementary extremal joins of the points

$$(9) \qquad A_1, z_1^1, \ldots, z_\nu^1, A_2$$

are not all equal in J-length. At least one of these minimizing joins must be less in J-length than the subarc of $\zeta^\nu(z)$ which it replaces. Otherwise z^1 would be in the set ω, since the subcurves replaced have equal J-lengths. Thus Lemma 1 is true.

Definition of the f^ν-deformation Δ_e^η of f_β^ν. With e conditioned as in Lemma 1, f^ν is of class C^∞ in some neighborhood, relative to f_β^ν, of each ν-tuple

$$z \in \omega - U_e^\tau . \tag{10}$$

Moreover, f^ν is ordinary at z, by virtue of Theorem 21.1 since each J-normal critical ν-tuple is contained in U_e^τ. If η is a sufficiently small positive constant, there accordingly exists an η-neighborhood ω_η of ω relative to f_β^ν, such that f^ν is of class C^∞ and ordinary on the subset

$$\omega_\eta - U_e^\tau \qquad (\text{of } f_\beta^\nu) . \tag{11}$$

To define Δ_e^η, an f^ν-deformation Δ_1 on f_β^ν of the set (11) will first be defined. Use will be made of trajectories on f_β^ν orthogonal to nonsingular level manifolds L of f^ν. At a ν-tuple $z = (z_1, \ldots, z_\nu)$ in the domain (11), the level manifold of f^ν meeting z, say L^z, in nonsingular. A trajectory, say T^z, with initial ν-tuple z, orthogonal to level manifolds of f^ν, will be parameterized by R-length s measured from z in the sense of decreasing f^ν. If $b > 0$ is sufficiently small, there exists a trajectory T^z of R-length b orthogonal to L^z at each ν-tuple z prescribed in (11). For b sufficiently small a ν-tuple on T^z with parameter s varies continuously in f_β^ν with z and its parameter $s \in [0, b]$.

Definition of Δ_1. Under Δ_1 a ν-tuple z in (11) will be replaced at the time t by the ν-tuple on T^z with parameter $s = tb$. For $b > 0$ sufficiently small, Δ_1 will be an f^ν-deformation on f^ν_β of the domain (11). Before defining Δ^η_e a second f-deformation Δ_2 of (11) is required.

Definition of Δ_2. Under Δ_2, ν-tuples in $\omega_{\eta/2} - U^\tau_e$ are deformed as under Δ_1. For $0 \leq \hat{t} \leq 1$, ν-tuples in (11) at an R-distance $\eta - \frac{\hat{t}\eta}{2}$ from ω are deformed as under Δ_1 until the time $\hat{t}$ is reached and held fast thereafter. So defined Δ_2 is a continuous f^ν-deformation of the set (11) on f^ν_β. Δ_2 is an effective f^ν-deformation of each ν-tuple $z \in \omega - C\ell U^\tau_e$.

Completion of definition of Δ^η_e. The ν-tuples in $\omega_\eta \cap U^\tau_e$, shall be held fast, while ν-tuples in $\omega_\eta - U^\tau_{2e}$ shall be deformed as under Δ_2. For $0 \leq \hat{t} \leq 1$ ν-tuples in $\omega_\eta \cap U^\tau_{2e}$ at an R-distance $e + \hat{t}e$ from U^τ shall be deformed as under Δ_2 until the time $\hat{t}$ is reached and held fast thereafter. The deformation of ω_η, so defined, is continuous and leaves ν-tuples on the boundary of ω_η in f^ν_β fixed. The definition of Δ^η_e is completed by requiring that ν-tuples in $\underset{\sim}{C}\omega_\eta$ be fixed.

Lemma 2 follows readily.

Completion of proof of Theorem 1. By virtue of Lemmas 1 and 2 a product f^ν-deformation

$$\Pi = \Delta^\eta_e G_e \tag{12}$$

of f^ν_β on itself is well-defined. Π leaves ν-tuples in U^τ_e fixed. We shall prove the following:

(i) If κ is a sufficiently small positive constant, then for ν-tuples $z \in \underset{\sim}{C} U^\tau_{2e}$ the final image z^1 of z under Π is such that

(13) $$f^{\nu}(z) - f^{\nu}(z^1) > \kappa .$$

A remark will be recorded before the proof proper of (i) begins.

Remark. Suitably parameterized, the trajectory of z under G_e, followed by the trajectory of $G_e(z, 1)$ under Δ_e^{η} is the trajectory of z under Π.

Proof of (i). A ν-tuple $z \in \underset{\sim}{C} U_{2e}^{\tau}$ belongs to one of the two following sets of ν-tuples in f_{β}^{ν}:

$$S_1 = \{ z \in \underset{\sim}{C} U_{2e}^{\tau} \mid G_e(z, 1) \not\in \omega\}$$

$$S_2 = \{ z \in \underset{\sim}{C} U_{2e}^{\tau} \mid G_e(z, 1) \in \omega\} .$$

If $z \in S_1$ the left member of (13) is positive by Lemma 1. If $z \in S_2$ the left member of (13) is positive by Lemma 2 and our Remark. Since the left member varies continuously with z and the set $\underset{\sim}{C} U_{2e}^{\tau}$ is compact, (13) follows.

Theorem 1 will follow from statement (ii).

(ii) Corresponding to a value $c < b_r$ such that $b_r - c$ is sufficiently small there exists an integer N so large that the product deformation Π^N deforms f_{β}^{ν} on itself into $(\tau_r)_{2e} \cup f_c^{\nu}$.

A consequence of Conditions K on e will be used in the proof of (ii). There exists a constant $c_0 \in (b_{r-1}, b_r)$ such that

(14) $$U_{2e}^{\tau} \subset (\tau_r)_{2e} \cup f_{c_0}^{\nu} .$$

<u>Proof of</u> (ii). For $z \in f^{\nu}_{\beta}$, let z^N be the final image of z under Π^N. Consider the countable set

$$z^1, z^2, z^3, \ldots \tag{15}$$

of ν-tuples and let $\hat{z}$ be a limit ν-tuple of the set (15). It follows from (i) that $\hat{z}$ is in the open set U^{τ}_{2e}. Otherwise it would follow from (i) that the relation (13), with κ replaced by $\kappa/2$, would hold for infinitely many of the ν-tuples (15). This is impossible since values of f^{ν} are bounded below. It follows then from (14) that

$$\hat{z} \in (\tau_r)_{2e} \cup f^{\nu}_{c_0} \ . \tag{16}$$

Corresponding to a prescribed ν-tuple $z \in f^{\nu}_{\beta}$ there then exists a value $c(z) \in (b_{r-1}, b_r)$ and an integer $n(z)$ so large that

$$z^{n(z)} \in (\tau_r)_{2e} \cup f^{\nu}_{c(z)} \ . \tag{17}$$

Let c' be an admissible value of c slightly larger than $c(z)$. The neighborhood $(\tau_r)_{2e}$ is open in f^{ν}_{β}. If then (17) holds for z, it also holds with $c(z)$ replaced by c' and z replaced by any ν-tuple in a sufficiently small neighborhood of z in f^{ν}_{β}. Since f^{ν}_{β} is compact, a finite set of such neighborhoods covers f^{ν}_{β}. Statement (ii) follows.

<u>Theorem</u> 1 <u>is implied by</u> (ii) <u>since</u> Π <u>leaves</u> ν-<u>tuples in</u> $(\tau_r)_e$ <u>fixed</u>.

Part II. Proof of Traction Theorem Ω_r. The final image of f_β^ν under the f^ν-deformation θ_e of Theorem 1 is a set

(18) $$A \subset (\tau_r)_{2e} \cup f_c^\nu \qquad (c \in (b_{r-1}, b_r))$$

of ν-tuples of f_β^ν. According to Definition 25.4, if e is sufficiently small the "broken extremal projection" π of $(\tau_r)_{3e}$ into $((M))_{\gamma_r}^\nu$ is well-defined. We then set

(19) $$N_e = \pi(\tau_r)_{2e}$$

and verify the following lemma.

Lemma 3. *For $e > 0$ sufficiently small, there exists an f^ν-deformation H_e of f_β^ν on itself that deforms $(\tau_r)_{2e}$ into N_e, leaves $N_e \cap (\tau_r)_{2e}$ fixed and deforms ν-tuples of f_β^ν into*

(20) $$N_e \cup f_c^\nu \qquad \text{(for some } c \in (b_{r-1}, b_r))$$

It should be noted that $(\tau_r)_{2e}$ is an open subset of $(M_n)^\nu$ and hence $n\nu$-dimensional while N_e is an open subset of $((M))_{\gamma_r}^\nu$ and hence $(n-1)\nu$-dimensional. We have set $\mu = (n-1)\nu$.

Definition of H_e. Before defining H_e an auxiliary f^ν-deformation h_e of $(\tau_r)_{3e}$ on f_β^ν will be defined.

Under h_e a ν-tuple $z \in (\tau_r)_{3e}$ shall have a replacement

(21) $$z^t = (z_1^t, \ldots, z_\nu^t) \qquad (0 \le t \le 1)$$

at the time t such that z_i^t moves along $\zeta^\nu(z)$ from z_i to $\pi_i(z)$, the i-th vertex of $\pi(z)$. This movement shall be such that the J-length of $\zeta^\nu(z)$ between z_i and z_i^t changes at a constant rate. h_e is a well-defined f^ν-deformation of $(\tau_r)_{3e}$ if e is sufficiently small.

Under H_e, ν-tuples in $(\tau_r)_{2e}$ are deformed as under h_e while ν-tuples in $\underset{\sim}{C}(\tau_r)_{3e}$ are fixed. For $0 \le \hat{t} \le 1$, a ν-tuple in $(\tau_r)_{3e}$ at an R-distance $3e - \hat{t}e$ from τ_r is deformed as under h_e until t equals $\hat{t}$ and held fast thereafter.

Lemma 3 follows.

Theorem 1 and Lemma 3 have a corollary which will replace Theorem 1, as an aid in proving the Traction Theorem. We set $P_e = H_e \theta_e$.

Corollary 1. *For* e *sufficiently small* P_e *is an* f^ν*-deformation of* f_β^ν *on itself, that leaves* ν*-tuples in* $N_e \cap (\tau_r)_e$ *fixed and for some* $c \in (b_{r-1}, b_r)$ *deforms* f_β^ν *into*

(22) $$N_e \cup f_c^\nu .$$

Completion of the proof of the Traction Theorem. The diff Φ_r introduced in (4) is such that $\Phi_r(D_\rho^\mu)$ is well-defined for arbitrarily small values of ρ and is a topological μ-ball that serves as a neighborhood of τ_r relative to $((M))_{\gamma_r}^\nu$. For any such value of ρ and for $e_r < \rho$ set

(23) $$\hat{\lambda}_r = \Phi(D^{\mu}_{\rho}), \quad \lambda_r = \Phi(D^{\mu}_{e_r}) .$$

Fixing ρ and hence $\hat{\lambda}_r$, let e and e_r be chosen in the order written so that

(24) $$\hat{\lambda}_r \supset N_e \supset \lambda_r \subset (\tau_r)_e .$$

The value e_r is arbitrarily small.

An f^{ν}-traction D_0 will now be characterized in Lemma 4. The product deformation $D_0 P_e$ will then serve as the desired traction T_r.

Lemma 4. *For any* $c < b_r$ *with* $b_r - c$ *sufficiently small, there exists a traction* D_0 *of* $\hat{\lambda}_r \cup f^{\nu}_c$ *into* $\lambda_r \cup f^{\nu}_c$.

Landis-Morse [1] Lemma 6.1 reads as follows:

(α) *If* $c < a$ *and* $a - c$ *is sufficiently small, there exists an* F-*traction of* $\Lambda_{\sigma} \cup F_c$ *into* $\hat{\Lambda}_{\sigma} \cup F_c$.

Lemma 4 will follow from (α) on making the following replacements:

(a_1) F and its domain F_{β}, by f^{ν} and its domain f^{ν}_{β};

(a_2) The critical point σ of F, by the critical ν-tuple τ_r of f^{ν};

(a_3) The critical value $a = F(\sigma)$, by the critical value $b_r = f^{\nu}(\tau_r)$;

(a_4) F_c with $c < F(\sigma)$, by f^{ν}_c with $c < f^{\nu}(\tau_r)$;

(a_5) Λ_{σ} and $\hat{\Lambda}_{\sigma}$, by $\hat{\lambda}_r$ and λ_r, respectively.

Subject to (24), statements (i) and (ii) will be proved.

(i) $D_0 P_e$ *deforms** f^{ν}_{β} *on itself into* $\lambda_r \cup f^{\nu}_{c_1}$ *for some* $c_1 \in (b_{r-1}, b_r)$ *and hence into* $\lambda_r \cup f^{\nu}_c$ *for every* $c \in (c_1, b_r)$.

---

*All deformations in this proof are f^{ν}-deformations.

(ii) D_0P_e deforms $\lambda_r \cup f^\nu_c$ on itself for some $c \in (c_1, b_r)$.

Proof of (i). By Corollary 1, P_e deforms f^ν_β on itself into $N_e \cup f^\nu_{c_0}$ for some $c_0 \in (b_{r-1}, b_r)$ and, since $\hat{\lambda}_r \supset N_e$ by (24), into $\hat{\lambda}_r \cup f^\nu_{c_0}$. By Lemma 4, D_0P_e then deforms f^ν_β on itself into $\lambda_r \cup f^\nu_{c_1}$ for some $c_1 \in (c_0, b_r)$. Hence (i) is true.

Proof of (ii). As an f^ν-deformation, P_e deforms f^ν_c on itself for any $c \in (b_{r-1}, b_r)$. P_e leaves λ_r pointwise fixed, since $N_e \cap (\tau_r)_e$ is so fixed under P_e and $\lambda_r \subset (N_e \cap (\tau_r)_e)$ by (24). The traction D_0 of Lemma 4 deforms $\lambda_r \cup f^\nu_c$ on itself for some $c \in (c_1, b_r)$. From (i) and Corollary 1 it follows that for this c, D_0P_e deforms $\lambda_r \cup f^\nu_c$ on itself. Thus (ii) is true.

It follows from (i) and (ii) that D_0P_e will serve as the traction T_r of our Traction Theorem Ω_r.

The tractions $T_r, \ldots, T_1$. According to the program outlined in §26 following (26.16), the above Theorem Ω_r is the first of r theorems of which the ith follow. In this theorem we refer to the neighborhood

$$\lambda_i = \Phi_i(D^\mu_{e_i}) \qquad \text{(rel. to } ((M))^\nu_{\gamma_i}) \tag{2.5}$$

of the ν-tuple τ_i. λ_i is introduced in Lemma 6.1.

Traction Theorem Ω_i. $i = 1, 2, \ldots, r$. If c_i is given as in (26.16), then for any sufficiently small $e_i > 0$ and for some $c_{i-1} \in (b_{i-1}, b_i)$ there exists an f^ν-traction T_i of $f^\nu_{c_i}$ into $\lambda_i \cup f^\nu_{c_{i-1}}$.

The proof of this theorem follows from the above proof of Theorem Ω_r on replacing r by i. The proof makes use of values of f^ν restricted to $f^\nu_{c_i}$. Theorem Ω_i leads to a choice of c_{i-1}.

<u>The traction</u> T_0. It remains to prove the following theorem. In this theorem we refer to the neighborhood

$$\lambda_0 = \Phi_0(D^\mu_{e_1}) \qquad \text{(rel. to } ((M))^\nu_{\gamma_0}) \tag{2.6}$$

of the ν-tuple τ_0. See (26.18) of Lemma 26.1.

<u>Traction Theorem</u> Ω_0. <u>If</u> $c_0 \in (b_0, b_1)$, <u>as in</u> (26.16), <u>then for any sufficiently small</u> $e_0 > 0$ <u>there exists an</u> f^ν-<u>traction</u> T_0 <u>of</u> $f^\nu_{c_0}$ <u>into</u> λ_0 <u>of</u> (2.6).

A proof of Theorem Ω_0 follows the lines of the proof of Theorem Ω_r, drastically modified. Simplifications arise because sets f^ν_c for which $c < b_0$ are empty.

A review of the proof of Theorem 1 of Part II leads to the following modification.

<u>Theorem</u> 1a. <u>If</u> $e > 0$ <u>is sufficiently small there exists an</u> f^ν-<u>deformation of</u> $f^\nu_{c_0}$ <u>on itself which leaves</u> ν-<u>tuples in</u> $(\tau_0)_e$ <u>fixed and deforms</u> $f^\nu_{c_0}$ <u>into</u> $(\tau_0)_{2e}$.

In reviewing the proof of Theorem 1 one replaces r by 0 and f^ν_β by $f^\nu_{c_0}$. With this change $U^\tau = \tau_0$ and $U^\tau_e = (\tau_0)_e$. Statements in the proof of Theorem 1 so interpreted, remain valid up to and including statement (i). Statement (i), with U^τ_{2e} replaced by $(\tau_0)_{2e}$, implies Theorem 1a.

One modifies the part of the proof of Theorem Ω_r contained in Part II as follows.

One replaces $N_e = \pi(\tau_r)_{2e}$ in (1.9) by $\dot{N}_e = \pi(\tau_0)_{2e}$ and Lemma 3 by the following.

__Lemma__ 3a. __For__ $e > 0$ __sufficiently small, there exists an__ f^{ν}__-deformation__ $\dot{H}_e$ __of__ $f^{\nu}_{c_0}$ __on itself that deforms__ $(\tau_0)_{2e}$ __into__ $\dot{N}_e$, __leaves__ $\dot{N}_e \cap (\tau_0)_{2e}$ __fixed and deforms__ $f^{\nu}_{c_0}$ __into__ $\dot{N}_e$.

The proof of Lemma 3a is similar to the proof of Lemma 3. The deformation $\dot{\theta}_e$ of Theorem 1a and $\dot{H}_e$ of Lemma 3a lead to the product deformation $\dot{P}_e = \dot{H}_e \dot{\theta}_e$ and to the following corollary of Theorem 1a and Lemma 3a.

__Corollary__ 1a. __For__ e __sufficiently small__ $\dot{P}_e$ __is an__ f^{ν}__-deformation of__ $f^{\nu}_{c_0}$ __on itself that leaves__ ν__-tuples of__ $N_e \cap (\tau_0)_e$ __fixed and deforms__ $f^{\nu}_{c_0}$ __into__ $\dot{N}_e$.

__Completion of proof of Traction Theorem__ Ω_0. For arbitrarily small positive constants $e_0 < \rho$ we follow (2.3) in setting

$$\hat{\lambda}_0 = \Phi_0(D^{\mu}_{\rho}), \quad \lambda_0 = \Phi_0(D^{\mu}_{e_0}) . \tag{2.6}$$

Fixing ρ and hence $\hat{\lambda}_0$, let e and e_0 be chosen in the order written so that

$$\hat{\lambda}_0 \supset \dot{N}_e \supset \lambda_0 \subset (\tau_0)_e . \tag{2.7}$$

Lemma 4 is replaced by a trivial lemma.

Lemma 4a. There exists an f^{ν}-traction $\dot{D}_0$ of $\hat{\lambda}_0$ into λ_0.

Statements (i) and (ii) are replaced by the following. Subject to the conditions (2.7) on e, e_0 and λ_0:

(i)a $\dot{D}_0\dot{P}_e$ deforms f_c^{ν} on itself into λ_0,

(ii)a $\dot{D}_0\dot{P}_e$ deforms λ_0 on itself.

The f^{ν}-deformation $\dot{D}_0\dot{P}_e$ will thus serve as the traction T_0 of Traction Theorem Ω_0.

A corollary of Traction Theorem Ω_0 follows.

Corollary 2. If $b_0 < c_0 < b_1$ the connectivities of $f_{c_0}^{\nu}$ all vanish except that $R_0 = 1$.

The corollary follows from the fact that λ_0 *is a topological* μ-ball and that $f_{c_0}^{\nu}$ is deformable on itself into λ_0.

Bibliography

Almgren, F. J., Jr.

1. The homotopy groups of the integral cycle groups. *Topology* **1** (1962), 257-299.

Atiyah, M. and Singer, I. M.

1. The index of elliptic operators I, II, III. *Ann. of Math.* **87**(1968), 484-604.

Berger, M. S.

1. New applications of the calculus of variations in the large to non-linear elasticity. *Comm. Math. Phys.* **35**(1974), 141-150.

Birkhoff, G. D.

1. The restricted problem of three bodies. *Rend. Circ. Mat. Palermo* **39**(1915), 1-70.
2. "Dynamical Systems." *Colloquium Publications*, Vol. 9. American Mathematical Society, Providence, R.I., 1966. (Rev. ed.)

Bliss, G. A.

1. *Lectures on the Calculus of Variations.* University of Chicago Press, Chicago, 1946.

Bôcher, M.

1. *Introduction to Higher Algebra.* Macmillan, New York, 1938.
2. *Leçons sur les Méthodes de Sturm.* Gauthier-Villars, Paris, 1917.

Bolza, O.

1. *Vorlesungen über Variationsrechnung.* B. G. Teubner, Leipzig and Berlin, 1909.

Bibliography 2.

Borsuk, K.

1. Sur les rétractes. Fund Math. 17(1931), 152-170.

Bott, R.

1. "Morse Theory and its Application to Homotopy Theory." Univ. of Bonn, Bonn, 1960.

Bourbaki, N.

1. Elements of Mathematics. General Topology. Part 1. Addison-Wesley Publishing Co., Reading, Mass. 1966.

Cairns, S. S. and Morse, M.

1. Fréchet numbers and geodesics on surfaces. To be published.

Carathéodory, C.

1. Variationsrechnung und Partielle Differentialgleichungen erster Ordnung. B. G. Teubner, Leipzig and Berlin, 1935.

Cerf, J.

1. La stratification naturelle des espaces de fonctions différentiables réeles et le théorème de la pseudo-isotopie. Inst. Hautes Études Sci. Publ. Math. No. 39, Paris, France, 1970.

Chern, S. -S.

1. Minimal surfaces in an Euclidean space of N dimensions. "Differential and Combinatorial Topology," pp. 187-198, Princeton University Press, Princeton, N. J., 1965.

Eilenberg, S.

1. Singular homology theory. Ann. of Math. 45(1944), 407-447.

Eilenberg, S., and Steenrod, N.

1. Foundations of Algebraic Topology. Princeton University Press, Princeton, N. J., 1952.

Eisenhart, L. P.

1. Riemannian geometry. Princeton University Press, Princeton, N. J., 1926.

Èl'sgol'c, L. E.

1. Calculus of Variations. Pergamon Press, London, 1961.

Federer, H.

1. Geometric measure theory. Springer, Berlin, 1969.

Flaschel, P. and Klingenberg, W.

1. Riemannsche Hilbertmannigfaltigkeiten. Periodische Geodätische. Lecture Notes in Mathematics, No. 282. Springer, Berlin, 1972.

Frankel, T.

1. Critical submanifolds of the classical groups and Stiefel manifolds. "Differential and Combinatorial Topology," pp. 37-53. Princeton University Press, Princeton, N. J., 1965.

Fréchet, M.

1. Sur quelques points du calcul fonctionnel. Rend. Circ. Mat. Palermo **22**(1906), 1-74.

Frolob, S., and Èl'sgol'c, L.

1. Limite inférieure pour le nombre des valeurs critiques d'une fonction, donnée sur une variété. Mat. Sb. **43**(1935), 637-643.

Gelfand, I. M. and Fomin, S. V.

1. Calculus of Variations (trans. by R. A. Silverman). Prentice-Hall, Englewood Cliffs, N. J., 1963.

Goldberg, S. I.

1. Curvature and Homology. Academic Press, New York, 1962.

Hadamard, J.

1. Leçons sur le calcul des variations, I. Paris, 1910.

Hermann, R.

1. Dynamical Systems and the Calculus of Variations. Academic Press, New York, 1968.
2. Focal points of closed submanifolds of Riemannian spaces. Nederl. Akad. Wetensch. Proc. Ser. A. 66(1963), 613-628.

Hestenes, M. R.

1. Calculus of Variations and Optimal Control Theory. John Wiley and Sons, New York, 1966.

Jordan, C.

1. Cours d'Analyse. Vol. 1, 3rd ed. Gauthier, Paris, 1909.

Kirby, R. C.

1. Stable homeomorphisms and the annulus conjecture. Ann. of Math. 89(1969), 575-582.

Kneser, A.

1. Lehrbuch der Variationsrecknung II. Braunschweig (1925).

Landis, D. and Morse, M.

1. Tractions in critical point theory. Rocky Mountain J. Math. 5(1975).

2. Geodesic joins and Fréchet curve classes. Rend. Circ. Mat. Palermo. (1) 8, Ser. VI (1975), 161-185.

Levi-Civita, T.

1. Sur l'écart géodesique. Math. Ann. 97(1926), 291-320.

Ljusternik, L. A.

1. The Topology of Calculus of Variations in the Large. Translations of Mathematical Monographs. Vol. 16. American Mathematical Society, Providence, R. I., 1966.

Ljusternik, L. and Snirel'man, L.

1. Topological methods in variational problems. Tr. Sci. Invest. Inst. Math. Mech. 11(1930).

Massey, W. S.

1. "Algebraic Topology: an Introduction." Harcourt Brace & World, New York, 1967.

Mazur, B.

1. Morse theory. "Differential and Combinatorial Topology," pp. 145-165. Princeton Univ. Press, Princeton, N. J., 1965.

Milnor, J.

1. Morse Theory. Princeton University Press, Princeton, N. J. 1969.

2. On manifolds homeomorphic to the 7-sphere. Ann. of Math. 64 (1956), 399-405.

Morse, M.

1. Variational Analysis. John Wiley and Sons, New York, 1973.

2. "The Calculus of Variations in the Large," Colloquium Publications, Vol. 18, 4th Printing, American Mathematical Society, Providence, R. I., 1965.

3. Fréchet curve classes. J. Math. Pures Appl. **53**(1974), 291-298.

4. Topologically nondegenerate functions. Fund. Math. **88**(1975), 17-52.

5. Functional topology and abstract variational theory. Mémorial des Sci. Math **92**(1939), 1-79.

6. Singleton critical values. Bull. Inst. Math. Academia Sinica **2** (1974), 1-17.

7. Connectivities R_i of Fréchet spaces in variational topology. Proc. Nat. Acad. Sci., U.S.A., **72**, No. 6 (1975), 2069-2070.

Morse, M. and Cairns, S. S.

1. Critical Point Theory in Global Analysis and Differential Topology. Academic Press, New York, 1969.

Myers, S. B.

1. Riemannian Manifolds in the Large. Duke Math. J., **1**(1935), 34-49.

Okubo, T.

1. On the curvature of a reductive homogeneous S/G associated to a compact Lie group. Preprint.

Palais, R.

1. Morse Theory on Hilbert manifolds. Topology **2**(1963), 299-340.

Palais, R. and Smale, S.

1. A generalized Morse Theory. Bull. Amer. Math. Soc. **70**(1964), 165-171.

Postnikov, M. M.

1. "Introduction to Morse Theory" (Russian). Izdat. "Nauka," Moscow, 1971.

Rauch, H. E.

1. *Geodesics and curvature in differential geometry in the large.* Yeshiva University, New York, 1959.

Rothe, E. H.

1. Critical point theory in Hilbert space under general boundary conditions. *J. Math. Anal. Appl.* **11**(1965), 357-409.

Sard, A.

1. The measure of the critical values of differentiable maps. *Bull. Amer. Math. Soc.* **48**(1942), 883-890.

Schoenberg, I. J.

1. Some applications of the calculus of variations to Riemannian geometry. *Ann. of Math.* **33**(1932), 485-495.

Seifert, H.

1. Minimalflächen von vorgegebener topologischer Gestalt. *S.-B. Heidelberger Akad. Wiss. Math.-Natur. Kl.*, 1974, 1. Abhandlung.

Seifert, H. and Threlfall, W.

1. *Variationsrechnung im Grossen.* Chelsea, New York, 1951.

Signorini, A.

1. Esistenza di un'estremale chiusa dentro un contorno di Whittaker. *Rend. Circ. Mat. Palermo* **33**(1912), 187-193.

Bibliography 8.

Smale, S.

1. Morse Theory and a nonlinear generalization of the Dirichlet Problem. Ann. of Math. **80**(1964), 382-396.

Takeuchi, M.

1. Cell decompositions and Morse equalities on certain symmetric spaces. J. Fac. Sci. Univ. Tokyo Sect. I, **12**(1965), Part 1, 81-192.

Thom, R.

1. Qualques propriétées globales des variétés différentielles. Comment. Math. Helv. **29**(1954), 17-85.

Tromba, A.

1. Morse Lemma in Banach Spaces. Proc. Amer. Math. Soc. **34** (1972), 396-402.

Uhlenbeck, K.

1. The Morse index theorem in Hilbert space. J. Differential Geometry **8**(1973), 555-564.

Yano, K. and Bochner, S.

1. Curvature and Betti numbers. Princeton, 1953.

INDEX OF TERMS

Library of Congress Cataloging in Publication Data

Morse, Marston, 1892-
Global variational analysis.

(Mathematical notes series ; no. 16)
Bibliography: p.
Includes index.
1. Differentiable manifolds. 2. Global analysis (Mathematics) 3. Calculus of variations. I. Title. II. Series: Mathematical notes (Princeton, N. J.) ; no. 16.
QA614.3.M67 515'.64 76-836
ISBN 0-691-08077-1